W9-CZO-015

# THE SEARCH FOR
# A THEORY OF MATTER

# THE HISTORY OF SCIENCE

Prepared under the general
editorship of Daniel A. Greenberg

# THE
# SEARCH FOR
# A THEORY
# OF MATTER

## by MENDEL SACHS

McGRAW-HILL BOOK COMPANY

NEW YORK • TORONTO • LONDON • SYDNEY
ST. LOUIS • SAN FRANCISCO • MEXICO • PANAMA

## ACKNOWLEDGMENTS

I should like to thank Dr. Daniel A. Greenberg for encouraging me to write this book. I am also grateful to Miss Barbara Milbauer for her excellent help in the preparation of the manuscript. Last (but not least) I would like to thank my son, Danny, for his ideas in leading to most of the cartoon illustrations in the text.

## *To my parents, Samuel and Florence Sachs*

PICTURE CREDITS

We are grateful for permission to reproduce the following material:

Illustrations on pages 17, 18, 29, 31, 35, 37, 38, 44, 48, 53, 55, 66, 67, 69, 73, 75, 80, 81, 85, 89, 104, 109, 113, 126, 130, 150, 159, 162, 193, 195 courtesy The Burndy Library, Norwalk, Connecticut.

The illustration on page 71 courtesy Arthur H. Compton X RAY AND ELECTRONS, MacMillan, London, Ltd.

The illustration on page 115 courtesy Dartmouth College.

The illustration on page 128 courtesy John Thomson CONDUCTION OF ELECTRICITY THROUGH GASES, Langevin, The Annales de Chimie et de Physique, Mason et Cie.

The sketches on pages 20, 26, 42, 43, 45, 57, 58, 60, 62, 79, 93, 137, 155 were drawn by Anne Marie Jauss.

# CONTENTS

Man's search for an understanding of the material world that surrounds him has a history that can be described with a small number of themes. One of the important ones, which has been considered since the earliest recorded studies of matter, concerns the question of whether *discreteness* or *continuity* is a fundamental characteristic of matter. That is to say, if one should think of a piece of clay, for example, and imagine it to be divided up into finer and finer parts. Would a time be reached where there would be an indivisible "atom of clay"? Or might the division process go on indefinitely, in a continuous fashion, until there would be nothing left?

It is interesting that throughout the history of science, the question about discreteness or continuity as fundamental features of matter has persisted, with the two different viewpoints triumphing in different periods as added experimental evidence revealed more and more about the nature of matter. Such continual change in viewpoint has indeed provoked man to search deeper and deeper toward a more complete understanding of the world.

A very peculiar thing has happened in our own century. We have reached a stage in our knowledge about matter where it appears that both views—discreteness and continuity—must be adopted simultaneously. But is this not a paradox? Each of these concepts seem to be incompatible with the other! Nevertheless, the majority of contemporary scientists have accepted this view of the fundamental nature of matter—that it can be described in terms of an accumulation of *discrete bits* under some experimental conditions, and as *continuous waves* under other experimental conditions. This is referred to as the "wave-particle duality."

In this book, we will trace man's search for a theory of matter, with primary emphasis given to the development and conflict of ideas that have emerged in the 20th century. The essential role of the provoking question of discreteness versus continuity will be stressed and the attempts to fuse the quantum theory of elementary particles with the theory of relativity will be discussed in a qualitative fashion.

We start with a brief discussion of the attitude toward the nature of matter as it was taken in ancient Greece by Aristotle and Archimedes. We then proceed to the revolution of Galileo and Newton (in the 16th and 17th centuries) in which previous approaches to a theory of matter are fused and generalized. The continuous field concept of Michael Faraday (in the 19th century) to describe electric and magnetic phenomena is then discussed. The theoretical and experimental discovery of the atomic nature of the gaseous state of matter (as found by Dalton and Boltzmann) then bring us to the latter part of the 19th century.

Next, we discuss the experimental results that appeared in the latter part of the 19th century and the 20th century, that disagreed with the "classical theories." These eventually led to the simultaneous births in this century of two revolutions in physics—*the quantum theory and the theory of relativity.* The conceptual bases of these two schools will be discussed separately and they will then be compared. Finally, the present-day attempts to fuse these two approaches—their successes, failures and hopes for the future—will be discussed in detail.

After completing this account of man's search for a theory

of matter, I hope that the reader will become aware of the essential role that the conflict of ideas must play in contributing to progress in our understanding of the world. If this awareness will leave the reader with an uncomfortable feeling, let him recall Newton's comment that man is but a grain of sand on the edge of a vast ocean of knowledge. As a real grain of sand on the beach, man could choose to lie dormant and comfortable in the warm sun. Or, if he is sufficiently uncomfortable, he might choose to probe the ocean of knowledge that lies before him. When he follows the latter course, he will find that the intellectual excitement of discovery that will be gained is certainly worth his discomfort and his effort.

*Mendel Sachs*
*Williamsville, New York*

# Introduction

The history of man's study of the material world around him can be divided among several different major themes. One of the earliest and most important of these themes concerned what the world was made of. Scientists and philosophers used the word *matter* to refer to substances that make up physical objects. The largest physical object with which we are most familiar is the earth.

As far back as the ancient Greeks, the people who studied matter were divided between those who thought that matter was made of many tiny particles and those who thought that matter could not be divided into particles. The idea that matter was made of many tiny particles is called the *discrete theory* of matter. The other idea that matter could not be divided into particles is called the *continuous theory* of matter. For example, the first group of scientists would say that even though a table top appeared to be a solid piece of wood, it was really made of tiny particles. The second group would say that it really was solid.

As knowledge of the world has increased, first one theory and then the other has seemed to be the correct one. The fact that we have never been certain which theory is the right one has led man to more and more detailed studies of matter.

In our century some scientists have come to believe that both theories are valid. Other scientists believe that matter cannot be both discrete and continuous. They believe that matter is either discrete or continuous. The majority of present-day physicists accept the theory that matter can be described as an accumulation of discrete bits under some experimental conditions and also as continuous waves under other experimental conditions. This idea that matter can be both discrete and continuous is called the *wave–particle duality*.

Even when a new theory has come along, and problems appear to be solved, we nearly always find that the newest explanation raises new questions which, in turn, lead to questioning the very basis of the theory. Thus, once more, the search continues. It is important for us to realize that the whole basis of science is not to find "the answer" but always a "better answer."

Science is like a living organism. The human body, for example, is constantly replacing old cells with new ones. Biologists have discovered that all of the cells of the human body at any given time will be replaced in a period of seven years. Science, too, replaces many of its old ideas and in some cases presents completely different ones. Yet what most often happens is that we keep parts of theories and discard other parts.

Rather than think of this continual search for new ideas as fruitless, one might instead think of the scientists' quest for knowledge as an exciting journey into unexplored regions of our universe.

Our reaction to matter begins from the moment we are born. As we grow we add to our experience of the world around us. One of mankind's most exciting and fundamental activities is to gain a basic understanding of the world in which he lives.

Scientists of ancient Greece who were called natural philosophers made great advances in understanding their environment. In the post-Renaissance period when mathematicians developed analytical mathematics, man made the next great step toward understanding his universe. Mathematical analysis was an important development because it enabled scientists to express their ideas in an exact way with numbers and also predict future phenomena. For example, scientists could predict mathematically when and where a comet might appear. Equally as important, they could compare their prediction with the actual event.

In the nineteenth century, scientists found that their available theories of moving things were inadequate to explain all of the properties of the matter they observed. This was especially true of electricity and magnetism, where a new sort of theory had to be proposed in order to explain these phenomena. This new theory was based on the idea that the *continuous forces* exerted by matter on other matter, rather than the discrete quantities of matter, should be the starting point of any basic description of matter.

In the twentieth century, the discovery of the *theory of relativity* favored the idea of a theory called the *continuous field theory*, even though many people were reluctant to give up the idea of discreteness. At the present time scientists have tried to coordinate both views, since neither theory alone is adequate to explain observed phenomena. In over two thousand years of scientific research we still have not reached a satisfactory understanding of the properties of matter.

The laws of physics which describe the effects of external forces on material things comprise a branch of physics called *mechanics*. Mechanics is usually separated into two categories, statics and dynamics. These describe the situations in which matter is in equilibrium, or at rest (static), and in motion (dynamics).

Mechanics is one of mankind's oldest fields of study. This is so because even his earliest attempts to study the heavens—the stars and planets—were actually mechanics. The Greek philosopher Aristotle (384–322 B.C.) was using mechanics when he described the motion of heavenly bodies. He based his theory of mechanics upon the ideas of the Greek philosopher Plato (427?–347 B.C.) who believed that since the heavens were perfect, heavenly bodies would have to move in perfect circles. He also believed that the earth was the stationary center of the universe since man, the most perfect object in the universe, lived on earth. Thus the earth was located in a special position in the universe—its center. Aristotle, on the basis of Plato's theory, believed that heavenly bodies revolved around the earth in perfect circles.

Aristotle's theoretical approach reduced his dynamics to a static situation because no motion was involved; that is, the passage of time was not applied to the orbits of the heavenly bodies around the earth. Observations made by Copernicus reflected Aristotle's theory. Copernicus said that the sun was the center of the universe around which all the planets including the earth revolved. Later Kepler, using Tycho Brahe's astronomical observations, proved that the orbits were not circular but elliptical.

Another Greek scientist who had a great influence on the course of physics was Archimedes (287?–212 B.C.). His principle of the lever explained in quantitative (mathematical) form the practical knowledge known at that time, namely, that a heavy object could be lifted by exerting a small force. This is done by placing the heavier object at one end of the lever very close to a fulcrum point. By exerting a downward thrust at the other end of the bar, the weight could be easily lifted.

Archimedes also discovered the principle of buoyancy, supposedly while sitting in an overflowing bathtub. From watching the water he deduced that when a body with a particular density is immersed in a substance of a different greater density, the body will displace a volume of the second substance equal to its weight. In other words, when Archimedes sat partially immersed in the bathtub he did not displace a quantity of water equal to his own volume. He displaced a quantity of water equal to his own weight. Another example was Archimedes' testing of the metal of Hiero's crown. Supposedly he took a piece of gold of known weight and immersed it in a container of water. He then measured the weight of the gold in the water. Next he took a piece of metal from Hiero's crown, equal in weight and thought to be gold, and repeated the process. From the earlier discovery, Archimedes had deduced that the weight of a body, immersed in the water, must be diminished by a buoyant force that is the weight of its volume of water. Thus if the metal with the same weight as a piece of gold should weigh the same in water, it has the same volume—and thus the same density. It must therefore be gold.

Both Aristotle and Archimedes made enormous strides which

*1. Nicholas Copernicus (1473–1543)*

influenced the future course of physics. Aristotle introduced the art of reasoning from statement to necessary conclusion into physics, in order to construct a consistent and valid theory of matter. Archimedes' method complemented that of Aristotle, for while Aristotle's emphasis was on logical structure in a true theory of matter, Archimedes focused his attention on the quantitative behavior of matter—that is, on the basis of experiments.

*2. Archimedes (287 ?–212 B.C.)*

Of course, they each invoked the other's method to some extent. Aristotle did experiments to complement his studies, and Archimedes theorized to complement his investigations. They both recognized that the experimental *and* the theoretical aspects of research were absolutely necessary in order to reach a true understanding of the behavior of matter.

The theories of matter that were developed in the Greek era were studied and expanded in succeeding generations. Nevertheless, not much progress was made in regard to the fundamental properties of nature until the Renaissance period in Europe and the work of Galileo Galilei (1564–1642). Many people think of Galileo as the father of modern physics. This is not so only because of his own contributions to science, it is also because of the *method* of science that he practiced and inspired others to follow. In his own research Galileo did not, of course, start from scratch! He based his investigations upon discoveries of his predecessors since the Greek era. Galileo's research was indeed profoundly influenced by both the deductive method of Aristotle and the quantitative approach of Archimedes. However, his life's work demonstrated most poignantly that true progress in understanding could only be achieved by the combination of sifting the significant portions of the acquired knowledge to date and criticizing it. Galileo was also willing to reject a part of the accepted foundations of physical theory in his day if it was necessary to do so to arrive at a consistent, theoretical description of matter.

It was Galileo's method first to deduce the physical implications of hypotheses, and then to rely on experimen-

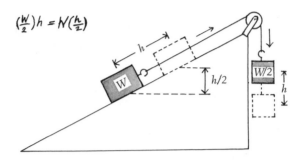

$$\left(\frac{W}{2}\right)h = N\left(\frac{h}{2}\right)$$

3. *Galileo's inclined plane*

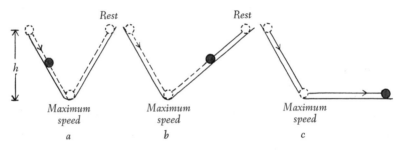

4. *Galileo's Principle of Inertia*

tation, rather than aesthetics, to verify the conclusion. To illustrate this, let us now follow Galileo's arguments and subsequent experiments which led him to conclude the mechanical properties of freely falling bodies.

Before experimenting on falling bodies, Galileo considered the following thought experiment. A weight on an inclined plane is held in position by a freely hanging weight that is connected to it by a pulley (Figure 3). He argued as follows: Let us say that the hanging weight is set into motion and allowed to descend by a certain amount. Clearly, the first weight will then move up the inclined plane through the same displacement. Galileo's intuition then led him to continue as follows: The equilibrium between these two bodies cannot be determined by their weights alone. It must also depend on their distance from the center of the earth. He then concluded: Since the second weight descended a certain vertical height, the first weight, in its motion up the inclined plane, must ascend a *vertical height equal to the height descended by the hanging weight multiplied by the ratio of weights of the hanging object and the block on the inclined plane.* Thus, the product of the weight times the *vertical distance* for the ascending weight is equal to the product for the hanging weight. These arguments were based solely on Galileo's intuition about the symmetry in this problem. The genius that was implicit in Galileo's theoretical conclusions, before any experiments were carried out, led the nineteenth-century physicist and philosopher Ernst Mach to make the following comment about him: "It is in this fact that Galileo discloses such scientific greatness: that he had the *intellectual audacity* to see, in a subject long before investigated, *more* than his predecessors had seen, and to trust his own perceptions."

Galileo's next line of thought was in relation to the *motion* of a freely falling body. Galileo deduced from his earlier consideration of the body at rest on an inclined plane that if this body should *freely* slide down the plane, it would have to increase its speed until a maximum speed was reached at the bottom. Now, if upon arrival at the bottom of the inclined plane the body should face another ascending plane, then the body would proceed up the second plane, gradually losing speed until it stopped. Furthermore, Galileo concluded, the maximum height reached on the second plane would be equal to the original height from which the body was released. This is because the initial and final positions have an equal amount of potential energy (that is, work capable of being produced). Galileo showed earlier that the amount of potential energy depended on the vertical displacement of the body. It followed from this line of thought that the maximum speed that could be acquired by the body in its fall depended only on the height from which it was released.

Finally, Galileo's instinct told him that the increase in speed with time (acceleration), as the body proceeds down the inclined plane, must be a constant. The average speed of a body that is uniformly accelerated is equal to one-half of the sum of the initial speed and final speed; and the distance traveled is equal to the average speed multiplied by the time of travel. He then deduced that the distance traveled depends on the square of the time of travel. Galileo was then able to test the validity of this formula by measuring the ratios of the times taken for bodies to slide down an inclined plane through different distances.

By using a water clock which he devised, Galileo showed experimentally that for any two distances down an inclined plane and their corresponding squares of times traveled, the ratios were constant and independent of the quantity of matter in motion. This remarkable result indicated that the rate of descent of a freely falling body is indeed independent of the intrinsic properties of the body itself! It implies, for example, that if a steel cannonball and a feather should be dropped simultaneously from the top of a building, they would land on the ground at precisely the same time—providing that they would fall in a vacuum. This assertion was incredible to the Renaissance man since it was so contrary to his "common sense." However, his "common sense" was based upon his experience with the world around him. Clearly, objects had never been seen falling freely in a vacuum. There was always air to impede the motion of the falling object.

Galileo and other early scientists worked under difficult conditions because their equipment was so crude that it was not always possible for them to obtain accurate results. For example, it was not possible in Galileo's day to produce a vacuum that would be good enough to test his conclusion about the falling cannonball and feather. In the actual experiment, of course, the greater air resistance to the feather's motion would certainly cause it to land much later than the cannonball. Nevertheless, Galileo's thought experiment on the motion of bodies sliding down an inclined plane led to a *general law of falling bodies* which predicted the outcome of an experiment that could be carried out in practice, in spite of the fact that it was not possible to obtain a vacuum. It is true, of course, that one could reject the idea that a general law must exist, and be content with a different formula for each different experiment. In that case, it is not necessary to accept the idea that the motion of a body sliding down an inclined plane is related to the motion of bodies that are dropped from the top of a building. However, it was understood, even in Galileo's day, that genuine progress has been achieved throughout the history of science only when studies were directed toward the construction of a general theory—an overall set of laws that could be proven in particular cases.

# 3.

# Law of inertia and the principle of continuity

Galileo *deduced* the law of inertia: A body at rest, or in motion with constant speed, will maintain this state forever, unless it is acted upon by some external force. His conclusion was based on analysis of the following thought experiment. Consider the inclined plane experiment discussed previously, in which a body rolls down one inclined plane and up another. Galileo concluded that if the body starts to roll downward from a vertical height (provided there is no friction to impede the motion), it must come to rest on the upward slope at the same vertical height from which it descended.

Let us now consider the motion of the body on the upward slope as a function of the inclination of the slope from the horizontal. As we have seen earlier, the maximum speed of the body is acquired at the bottom of the first slope. As the body proceeds on its upward journey, it will decrease its speed until it stops at a certain vertical height—the same height from which it was originally released. Clearly, if we decrease the angle of the upward slope, the length of path that must be traveled by the body before it reaches its vertical height would be correspondingly increased (Figure 4). Consequently, as the angle of inclination of the upward slope is decreased, the time required before the body would come to rest is correspondingly increased. At the limit, where the angle of inclination becomes zero, that is, when the "upward slope" becomes horizontal, the maximum velocity of the body, which was achieved at the bottom of the downward slope, would not change at all. That is, the body would proceed along the horizontal surface with a constant velocity, forever, unless compelled to slow down or speed up by some external force. Again, Galileo's conclusion was incredible to the Renaissance man because it was so contrary to experience. Never had anyone seen a body in continual motion on a horizontal plane. However, he had to admit that in practice, there was always a frictional force present, no matter how small, so that experience did not refute Galileo's conclusion which was, after all, based on an idealized physical situation. Nevertheless, it was an idealization that led to a general law of nature that could be tested, and it proved to be correct. It was from the analysis of such a thought experiment that Galileo deduced the *law of inertia of matter*.

The law of inertia was not put into precise mathematical form until a generation later, when Isaac Newton (1642–1727) developed the mathematical analysis of continuously changing parameters, called differential calculus. Newton's *second law of motion* asserts that the rate of change of the speed of a body, called acceleration, is directly proportional to the external force that is imposed to produce this change. Thus the equation, force equals mass times acceleration, is a mathematical expression of Galileo's law of inertia. It was with this mathematical expression of Galileo's law that Newton defined the inertial mass of a body as a constant of proportionality between the acceleration of the body (the effect) and the applied force (the cause). Newton said that the inertial mass of a body is a measure of its resistance to a change in its state of rest or constant speed; the greater the inertial mass of a body, the greater must be the applied force to overcome its state of rest or constant speed.

Other far-reaching interpretations of inertial mass have been introduced; however, Newton's second law of motion has been

experimentally confirmed in a very great number of observations of moving matter. Thus it is certainly a valid empirical rule that must be included mathematically as a feature of any other general theory.

Let us look at a graphical representation of Newton's law. If one plots the force in the vertical direction and the acceleration that is induced in a mass in the horizontal direction, then one obtains a straight-line locus of points. At any point on this line, the ratio of a particular force to the corresponding value of the acceleration is equal to the mass. Different straight lines through the origin would then correspond to bodies with different masses. In Newton's day, it was not possible to measure "high-energy" deviations from the conventional linear law, and thus Newton's law was not in dispute. However, as experimental techniques have become progressively better in recent years, it has become necessary to explore the validity of the older Newtonian law in the extended region of the graph. In this region the quantities of energy and momentum that are transferred between interacting matter might be regions in which the

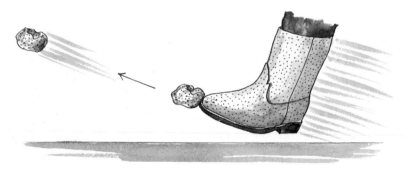

5. *When stone is kicked, does its direct contact with the boot cause it to move, or is the motion due to action-at-a-distance between the constituent atoms of the stone and the boot?*

Newtonian linear law might break down. We will see in later chapters why it became necessary to alter the theories of matter of Newton and Galileo, both in the conceptual and mathematical forms of their approach, in order to accommodate experimental discoveries of the twentieth century.

Finally, as a by-product of Galileo's analysis of the mechanics of moving bodies, he observed that if one should continuously vary the slopes of the uphill planes in which the body decelerates until it comes to rest, the effect would be to correspondingly change the deceleration in a continuous fashion. Galileo then concluded that the equations of motion of moving bodies must obey a principle of continuity. This principle may be stated as follows: Given any two values for a measured physical property of an object in motion (for example, speed, acceleration, or position) there must always exist an infinite number of potentially measurable values of this property between them. This is so no matter how close the two chosen values may be. *Differential calculus* is used to formulate this property of continuity. It was devised for specific application to moving bodies by Isaac Newton a generation after Galileo. Differential calculus was also developed in a slightly different fashion by Newton's contemporary, W. Leibniz (1646–1716).

While the principle of continuity appeared to Galileo and Newton to describe the motion of things, Newton did not take the view that things themselves should have a continuous description. Rather, he was a strong believer in the corpuscular ("atomic," or discrete) nature of the fundamental bits of matter that make up the universe. Let us now consider his line of thought in regard to this view.

# 4.

## The mechanical particle and action-at-a-distance

The natural philosophers of the past argued as follows: If the world is indeed composed of distinguishable bits of matter, then it is only common sense to conclude that their motion must be due to their effect on each other. The most natural way for matter to affect matter is by direct contact. For example, a pebble will remain where it is until I kick it. The instant my shoe touches the pebble, the pebble moves. Thus, the pebble moves by direct contact with my shoe. It is also obvious that the distance that the pebble will travel must depend on the force with which I kicked it and also on the weight of the pebble. Certainly if the pebble had been a boulder, not only would it not have moved from a forceful kick—I would probably have broken my toe in the process!

There is a famous legend that Isaac Newton was sitting under an apple tree one day when "plop!" an apple hit him on top of his head. Instead of jumping up with sudden anger, as most young men would do, Newton was intellectually aroused. Rather than shouting "ouch!" he asked a question: If action by contact is a basic law of nature, then what pushed the apple and made it fall on my head? Could it have been an unseen demon on the branch? Following this line of questioning, Newton may have thought: Why do I feel warm when the sun is high in

6. *Galileo (1564–1642)*

the sky and cool when it is on the horizon? What is it that is pushing the stars along their paths in the heavens? Are there larger demons doing this work? Because he did not believe in demons, it might have occurred to Newton that perhaps things influence each other without direct contact—that is, things may cause other things to move simply by virtue of their being there in space.

Clearly, the concepts of *action-by-contact* and *action-at-a-distance* could not be accepted simultaneously as basic laws of moving bodies, for they are laws that logically oppose each other. If things can only interact by contact, then they cannot interact at a distance! Newton felt he had to make a choice between these two concepts. Because of his intuition he favored the action-at-a-distance theory. This rule seemed realistic, since it covered experimental situations that were obviously action-at-a-distance, for example, the influence of the sun upon the earth. It also covered those situations which appeared to be governed by the action-by-contact principle. In those situations where direct contact seemed to be involved, one could still view the motion caused, in terms of action-at-a-distance between the bundles of atoms that in reality comprise the interacting bodies (Figure 5). It should be emphasized that Newton found it difficult to express what he actually meant by action-at-a-distance.

It is very difficult to understand the concept of mutual action-at-a-distance. As an illustration, a mechanical particle placed at an arbitrary location automatically "knows" the existence of all other mechanical particles, where they are located, and whether they are moving or stationary. Nevertheless, Newton was content at the time to accept the concept of action-at-a-distance as an empirical rule, even though he could not explain it. The idea was certainly compatible with the predictions of Newton's laws of nature and experience. Newton expressed the hope that the fundamental explanation of this concept might come at some future date.

### Universality of Gravitation

Returning to the apple incident, it is conceivable that Newton's thoughts might have proceeded as follows: Let us suppose that action-at-a-distance is the manner in which matter affects matter. Could the particles of matter in my head that interacted with the particles of matter in the apple have drawn the apple to me? Certainly not! I am quite sure that the apple would have dropped along the same path whether I was there or not. In fact, the apple would have followed the same path whether the tree were on sloping ground or on flat ground. I also know that the earth is a closed, spherical surface. If I sit under

*7. Sir Isaac Newton (1642–1727)*

an apple tree in London, or in Paris, or even in India, my experience would be the same. Suppose I drew a picture of all possible paths that apples could follow as they fall from trees at any point on the surface of the earth. It is clear that all of these paths must converge at a common point because of the spherical symmetry of the earth!

Newton might then have asked himself: Why is this so clear? Why is it common sense? He might have answered, if all of the apple trees on earth stay fixed in position while the earth is rotated about its axis, then there should be no change in the paths of falling apples. An apple from a tree in London or in Auckland, New Zealand, would still fall the same way, even if some demon should uproot them and interchange their positions. The only straight lines associated with the shape of the earth and which have this property are radial lines, that is, lines that go from the surface to the center of the earth. These lines are perpendicular to the surface of the earth and all converge at its center. Newton would then have concluded that the primary force that drew the apple to his head was actually a force directed along a radius of the earth. The apple was therefore moving toward the center of the earth; his head just happened to be in the way. Thus, Newton's conclusion completely agreed with Galileo's earlier deduction about the motion of bodies along inclined planes. The motion depends on the amount of attraction toward the center of the earth.

Newton argued that the *force of gravity* must depend on the mass of the earth. (Clearly, if one should imagine the mass of the earth to diminish continuously, its pull on other massive bodies would correspondingly diminish.) The force of gravity must certainly also depend on the mass of the apple (already deduced by Galileo); and now Newton reasoned that this force must depend on the distance between the apple and the center of the earth.

The next question that Newton asked was the following: In what way does the force of gravity depend on the distance between the interacting masses? The force must either increase or decrease as the distance between the interacting bodies is longer or shorter. For example, if two natural magnets are sufficiently separated, they soon stop having any effect on each other. Thus Newton concluded that it is common sense that the natural force of gravity must depend inversely on the separation of the interacting bodies, that is, the smaller the separation, the greater the force, and vice-versa.

Perhaps at this point, Newton remembered that the astronomical findings of Kepler and Galileo did actually indicate that the correct orbits for the planets is predictable if the mutual force between the sun and the planets is understood to have an inverse relationship, on the square of their respective separations. Thus, his strong belief in a general law of gravity led Newton to conclude that the nature of the force between the sun and the planets is no different than the nature of the force between the earth and the apple, or between any bodies that possess inertial mass. This force is called the force of gravity and depends on the product of the masses of the interacting bodies, on the inverse square of their separation between centers, and on no other variables. Finally, Newton concluded that if this mathematical relationship were to define the nature of the gravitational force—the same force that predicts the planetary orbits, the motion of the stars, the force that holds a man on the ground and causes an apple to fall to the earth—then the force of gravity must necessarily depend on a universal constant which determines precisely how strong this force is when massive bodies are separated by one distance or another. If the magnitude of the force of gravity could be determined from one particular experimental observation, the same constant must necessarily appear in the description of all other types of experiments that involved gravitational interaction between massive bodies.

While we sophisticates in the twentieth century might take the notion of a universal gravitational force for granted, its suggestion to mankind in the seventeenth century was a quite radical innovation in the thinking of that day. For it was very difficult for seventeenth-century man to imagine that the motion of the planets and the stars should have anything to do with why a man is held down to the floor of his house, or why a stone will roll down a hill, or why an apple drops down, rather than up, from a tree. Of course, the notion of searching for a general theory was not new. It was advocated by Newton's predecessor, Galileo; it was also sought from the earliest days by generations and generations of philosophers and scientists. Nevertheless, Galileo and Newton did manage to take this philosophical notion and apply it practically. They were able to state the idea of a general theory in a language that permitted precise, quantitative predictions which could be compared with the experimental facts. Thus, a giant step in the history of science was made by Galileo and Newton, not only by their verification of the universality of at least one of the natural forces, but more significantly, by their setting the stage for the method of theoretical and experimental physics in the search for a general theory of matter.

### Corpuscular Theory of Light

Newton's desire to use a general theory to describe the behavior of nature, rather than having a different theory for each separate physical phenomenon, led him to conclude that if one phase of nature (for example, gravitational force) is understandable in terms of grouped particles of matter that act on each other at a distance, then light, being a natural phenomenon, also must be composed of particles, that is, corpuscles of light. Newton, through his experiments, was aware of the fact that natural light shone through a prism will result in the breakdown of light into a spectrum, or a band of colors, of which natural light is composed.

Newton's experiments never yielded any evidence of light traveling in any path other than a straight line. His experience also seemed to show that the objects in its path would cast a sharp, geometric shadow. Thus Newton surmised that light could consist of particles, like the mechanical particles he had considered earlier. He further proposed that the different-colored light that emerged from a beam of white light, after being dispersed by a prism, was in fact caused by different types of light corpuscles moving at different speeds in such a medium, but all at the same speed in a vacuum. (These corpuscles constitute the mixture that we see as "white light.") In this way, Newton proposed a model that would correctly describe the effect of light dispersion as well as the observed, sharp, geometric shadows and the straight-line path for a light ray. He then felt confident that the general laws of nature that apply to the propagation of light were the same as those which apply to the bits of matter moving about in space.

8. A title page for one of Kepler's works shows Tycho Brahe, Ptolemy, Copernicus, Hipparchus, and Kepler at bottom right.

*The Law of Partial Pressures*

Believing that the material universe is composed of a very large number of microscopic atoms of matter, Newton and many of his predecessors thought that the bulk properties of matter must certainly be a function of the way in which the individual atoms of matter behave under various experimental conditions. In other words, the way in which the individual atoms that make up a piece of matter behave has a direct bearing on the way the whole piece of matter behaves. Any measurement of the properties of matter must then relate to the way large groups of atoms behave when they are subjected to particular experimental situations.

Some of the first quantitative experiments that further substantiated Newton's hypothesis about the atomic character of matter were carried out by the nineteenth-century British chemist John Dalton (1766–1844) in his studies of gaseous matter. Dalton's historic discovery, the *law of partial pressures,* asserts that the total pressure that is exerted by a gas on its surroundings is equal to the sum of partial pressures of the individual chemical constituents of the gas. If a gas is steadily released from a container, the pressure exerted by the gas on the walls of the container would correspondingly decrease. At the end, when only one atom of oxygen is left (this is an oxygen molecule) and one atom of nitrogen is left, then the total force per unit area—pressure—that is exerted on the walls of the container is the sum of the forces due to the single oxygen and nitrogen molecules contained in the walls. Should the oxygen molecule be removed from the container, the resulting pressure would then depend on the mass of the nitrogen molecule alone. Testing this experimentally, Dalton concluded that the law of partial pressures did in fact verify the atomic nature of matter. From these studies, Dalton also verified the more recent concept of discrete atomic weights.

9. *Michael Faraday (1791–1867)*

Finally, these investigations led to the prediction of the number of atoms that must be contained in a quantity of matter visible to the naked eye.

With these experimental facts, the mid-nineteenth-century scientist became convinced of the validity of the atomic nature of matter. Nevertheless, to make more accurate predictions about the properties of macroscopic matter, it became necessary to apply the laws of mechanics of Galileo and Newton to a fundamental analysis of gases, liquids, and solids—the three observable states of matter.

10. *The generator with which Faraday converted magnetism into electricity*

### The Role of Statistics in
### Classical Theories of Matter

If we accept the atomic nature of matter, it then follows that if a scientist knew the initial conditions of speed and location for all of the constituent atoms of some quantity of matter, then the laws of mechanics would permit him to predict, to any desired precision, the future course of all of these atoms. Such knowledge would provide an exact description of matter that could be subjected to any external conditions, provided that the nature of the forces was also known. Unfortunately, the scientist is not capable, through experimenting, of determining the initial conditions of location and speed for all constituent atoms. This is because the number of atoms in an observable quantity of matter has been found to be extremely large; it is something on the order of a million billion billions! Even if he should be able to do this impossible task, a further impossible task would remain: to solve the set of one million billion billion simultaneous equations that relate to the motion of these atoms. We see then that to verify the atomic model for matter, in fundamental terms, it becomes necessary to predict its physical properties without following the paths of individual atoms through time. To resolve this problem, it then becomes necessary to predict the average properties of a large group of atoms without knowing their individual speeds and locations.

One of the great achievements of nineteenth-century theoretical science was the development of *statistical mechanics*—a mathematical formalism that provides a method for finding the average properties of a large group of atoms without knowing the precise initial positions and speeds of each of its constituent elements. With the assumption that (1) the number of atoms in a given volume of gas is a constant, (2) the total energy for the ensemble of atoms is constant, and (3) that the entropy (a quantity which is expressed as a function of the temperature, pressure, and density of a system and is a measure of the amount of energy unavailable for work during a natural process) does not change, L. Boltzmann (1844–1906) was able to derive the most probable distribution of speeds for a large group of atoms. He was then able to derive all of the thermodynamic properties of a gas (pressure, specific heat, and entropy). He was able to show that the temperature of a gas is, in fact, nothing more than a representation of the average kinetic energy of its constituent atoms. Starting from these results, it then became possible to derive the conductivity of heat, the diffusion of temperature from a hot gas into a cold gas, and the viscosity of the gas. These predictions, which depended critically on the form of Boltzmann's equation, agreed with the observed properties of matter and thus established the atomic constitution for a gas.

Near the end of the nineteenth century, the same types of statistical calculations were applied to solid objects. Once again, the theoretical predictions agreed very well with the observed properties. Thus, both solids and gases appeared to have a firm theoretical explanation for very large numbers of atoms. Just as Western scientists since the Greek era had thought, the idea that matter is composed of discrete things appeared to be firmly established. So convincing was this evidence for the atomic model of matter, together with Galileo's and Newton's success in verifying a description of the constituent elements, that many of the leading scientists and philosophers of the late nineteenth and early twentieth centuries believed that mankind had finally completed its investigation of the fundamental properties of nature. They believed that Newton had indeed fully proved his corpuscular view of matter. It was believed that the action-at-a-distance concept and the mechanical laws of Newton that describe the basic constituents of matter, as well as the motion of the planets and the stars, was also firmly established. With this attitude, the only remaining obligation of physicists of future generations was to apply these "known principles" in order to discover more about the behavior of matter.

# 5

# The field concept of Faraday

In spite of all of this optimism about the atomic nature of matter, nineteenth-century studies on the electrical and magnetic properties of matter were not yet understandable in terms of discrete bits of matter. Although most of his colleagues adhered to the Newtonian view, Michael Faraday (1791–1867) understood electric and magnetic effects only in terms of *continuous fields*, which he considered to be the basic elements of a description of matter.

The term "field" refers here to an entity that is continuously distributed throughout all of space and time. To consider a field as the basic element from which to build a fundamental description of matter was, of course, quite contrary to the discrete theory. The discrete theory states that basic entities are restricted to discrete points or to very limited regions of space. To illustrate this difference, let us consider a drop of water as it falls to the ground. Upon impact the drop will spread out. Let us now apply the "field" concept to the physical property, "wetness." As the drop hits the ground, the wetness in the system can be measured only at one small confined region of space—the location of the water drop. After impact, on the other hand, the wetness could be experienced by a measuring device, say a toe, at any point on the ground. Thus, the situation before impact related to a dis-

*Before*　　　　　　　　　　　*After*

*11. Discreteness versus continuity*

crete form of wetness while the situation after impact relates to
a continuous form of wetness.

While the preceding discussion illustrates the difference
between the concepts of discreteness and continuity in one example, the basic laws of nature cannot accept both simultaneously, as the fundamental way to describe matter. Thus, the
basic elements of matter are discrete or they are continuous in
nature; they cannot be both at exactly the same time. Contrary
to the view of the atomists, Faraday proposed that the basic
elements that underlie the most general description of matter
are continuous entities, for example, light, gravity, electricity,
or magnetism. He took these to be the fields that relate directly
to the influence that one kind of matter would have on another.
This was indeed a revolutionary thought! Faraday was not trying
to say that one should deny the existence of the free, noninteracting bits of matter. Rather, he was proposing that it should
only be the manifestations of matter in influencing other matter
that must be considered as the basic entity. In this way, he was
forced to the conclusion that the basic entities that must be used

to describe matter are continuous rather than discrete. With this view, it no longer matters whether matter, in its noninteracting state, is discrete or not. It is only the continuous manifestation of matter in influencing other matter that counts. Faraday then proceeded to prove his point through some brilliant, experimental investigations of the electrical and magnetic properties of matter.

For example, consider the well-known experiment in which a bar magnet is placed beneath a sheet of paper, and iron filings are sprinkled on top of it. These little bits of iron will then be seen to assume a particular pattern on the paper. They seem to lie along curved lines that start at one end of the magnet and terminate at the other (Figure 12). In making this observation, Faraday asked the question: What is more fundamental in this observation, the iron filings or the "lines of force" that are aligning them? It seemed to him that it was the continuous field of influence of the magnet, which is being probed by the iron filings, that is the basic entity. It is this entity that represents the magnet. For without these "lines of force" the magnet would not be a

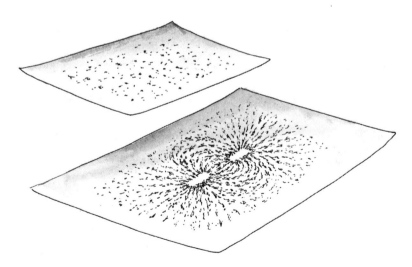

*12. An effect of a magnetic field of force*

13. *Electric lines of force and potential surfaces as drawn by James Clerk Maxwell in his book* ELEMENTARY TREATISE ON ELECTRICITY.

magnet! It is also true, he might have continued, that it would not really be meaningful to ask whether or not this field of influence is present if we did not have at our disposal a bag full of iron filings to probe it. That is, to have a field of force, there must be something in the system described that is subject to such a force. Nevertheless, it appears to me, Faraday would have concluded, that it is the continuous field itself, rather than the discrete bit of matter, that should be considered as the basic entity here.

Since Faraday believed that the continuous field of influence of matter should be considered as the basic entity, he felt that the electrical properties, as well as the magnetic properties of matter, should also have such a description. Analogous to the iron filing experiment in magnetism, Faraday established the field properties of electricity by studying the behavior of a salt solution in which one immerses a pair of oppositely charged electrical plates. By measuring the electrical potential inside of the solution as a function of position, a mapping of the continuous electric field intensity is thereby determined. The positive and negative ions in the salt solution flow along field lines just as the iron filings were forced by the magnetic field, in the previous example, to assume a particular spatial configuration (Figure 14). The electrical experiment just described is called electrolysis; it has a practical application in metal plating. If, for example, one should use a silver solution, then the positively charged silver ions in the liquid would flow toward (and plate) the negatively charged metal electrode.

Just as Galileo and Newton had made great strides in their studies because of their belief in a general theory, so Faraday also made gigantic progress because of his basic belief in the unity of science and in the existence of a general theory. After his studies of the electric and magnetic "fields," Faraday could not believe that these were really separate and totally independent of each other. Rather, he felt that these two seemingly different types of fields probably represent different manifestations of a single field that is subjected to different conditions. Thus he felt that the electric and magnetic fields could be derived from the same general laws. Faraday then proceeded to prove that this was the case; he set up an experiment in which a wire that is

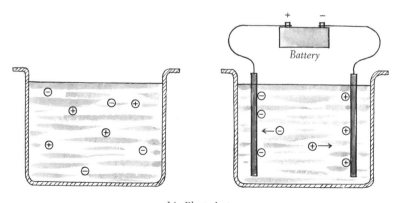

*14. Electrolysis*

capable of conducting electrical current is moved through a magnetic field. In doing so, he found that the wire became electrically polarized, thus causing electric current to flow through it. He then concluded that an electric field is equivalent to a magnetic field in motion, that is, the magnetic field was moving relative to the wire in this experiment. It should be possible to demonstrate the reverse process. Faraday also showed that an electric field in motion is equivalent to a magnetic field, which could be detected, for example, by a compass needle placed near the current carrying wire. Thus, Faraday made the very important discovery (Faraday's discovery of the relation between the electric and magnetic manifestations of matter in terms of relative motion of one field through the other led to the invention of the dynamo) that the electrical and magnetic properties of matter are really not separate phenomena; rather they were found to be a part of one general phenomenon which we now call *electromagnetism.*

Some of Faraday's last experiments were attempts to unify the electromagnetic fields of force with Newton's gravitational force field. Faraday's confidence in the existence of a general theory was very strong. Thus, he believed that Newton's action-at-a-distance approach was incompatible with the continuous field approach of electromagnetism. Both could not be correct because they were too dissimilar. It had to be one or the other. In his effort to find a general field theory, Faraday spent a great deal of time studying Newton's idea of action-at-a-distance. He finally concluded that the most appealing factors in Newton's theory were the mathematical calculations. It was these calculations that appeared to make Newton's theory valid.

Upon further reflection Faraday was struck with the thought that Newton's calculations might also be valid for the field theory. He set about trying to incorporate them into the field theory. His purpose in doing so was not only his belief that the general field theory was the more correct one, but also his strong conviction that the two theories could not be valid for an explanation at the same time. To prove the unification of electromagnetism with gravity in a *unified field theory*, Faraday proceeded to devise some very clever experiments in which combinations of electrical and gravitational effects were involved. Unfortunately, because of the very great difference between gravitational attraction and electrical attraction, it was not possible, with the equipment that was available to him, to demonstrate such a unity.

In addition, there were theoretical problems that stood in his way. To substantiate his claim that such a general field theory does exist, it became necessary for Faraday to answer the following question: If electromagnetic fields of force are interchangeable with gravitational fields of force, why is it that electromagnetic forces are found to be both attractive and repulsive, while gravitational forces have only been seen in nature to be attractive? While several possibilities have been proposed to answer this question of a unified field theory, both in Faraday's time and in the ensuing periods of twentieth-century physics, no satisfactory answer has yet been accepted. The question, nevertheless, is most crucial, one to be answered by a truly unified approach to physics.

## Maxwell's Field Equations

Following Faraday's introduction of the field concept into physics and his experimental confirmation of the electromagnetic field, it was necessary to state the results in mathematical language so that any inquirer could make precise predictions that could be checked with results of further experiments. James Clerk Maxwell (1831–1879) brilliantly achieved this task. Maxwell's *partial differential equations* are the means of expressing Faraday's field theory in mathematical language. The final form of these equations was suggested earlier in separate experiments of Faraday, Ampère (1775–1836), Biot (1774–1862), Savart (1791–

*15. James Clerk Maxwell (1831–1879)*

1841), and others. Maxwell described Faraday's theory by combining the mathematical description of various electric and magnetic effects into one set of equations. In addition, Maxwell's equations contained a very special property that strongly influenced Albert Einstein in his discovery of the theory of relativity in the early part of the twentieth century.

Let a scientist deduce Maxwell's equations from his own measurements of electric and magnetic effects that are made in his laboratory and according to the time measured with his own watch. Suppose that he should communicate with another scientist engaged in a similar project. The second scientist may be either stationary at some location or moving at a constant relative speed. For example, the second scientist could be on a boat at sea, moving at a constant relative speed, or he could be stationary, on the surface of the moon. After the two have come to an understanding of the language (that is, mathematical expressions), that each of them will use, they will make a discovery. They will discover that the different forms of Maxwell's equations that each deduced were precisely the same. Thus we see that this particular law of nature has the same form no matter how it is deduced. Such a set of equations is said to be invariant in form, or covariant. *Invariance* means a quantity is the same in all frames of reference moving at constant speeds with respect to each other. *Covariance* means that the quantity varies in accordance with a fixed mathematical relationship.

Before it is possible to prove the covariance of Maxwell's equations, it is, as noted, first necessary to translate the mathematical language of the two observers so that they can relate to each other. In particular, this refers to translating not only the spatial position, but also the time measure, making them relative to a specific observer; that is, the time as well as the space points are meaningful only with respect to the state of motion of the observer. Once this translation is accomplished the two scientists will find that they are in exact agreement with the form of Maxwell's equations. Maxwell's equations demonstrate *relativistic invariance* and led Einstein, in the early twentieth century, to revolutionize previous ideas and to formulate the basic laws for the behavior of matter. We will return to this concept in the chapters that follow.

In setting up Maxwell's field equations, the time coordinates always appear together with the spatial coordinates so that whenever the equations are translated from one language to another, the combined space and time coordinates translate into combinations of the four. It is necessary to treat the time coordinate as a dimension just like the space coordinate. In order to do this Maxwell had to introduce a constant into these equations. (In other words one cannot translate three horses into two cows in any language!) The dimension of the constant factor that converts the units of time (seconds) into the units of space (centimeters) is speed (centimeters per second). If one scientist is in Auckland, and one is in London, and they wish to compare the results of their experiments, they must agree on how to express the units of time in terms of the units of space. Thus, the constant speed that is used for this conversion must be the same for any scientist, irrespective of his location or motion—it must be a *universal constant.* It is usually denoted by the letter c.

When one examines the form of Maxwell's equations and observes how the constant speed enters these equations, we see that this constant relates to the speed of transmission of electric and magnetic effects between interacting matter that is electrically charged. It is interesting to note the sharp contrast between this approach and the corresponding feature of the action-at-a-distance approach. In the latter theory, one particle of matter "feels" the other, simply by virtue of its existence at a particular location in space. This is equivalent to saying that the information about the existence of one bit of matter transmits itself to the other with infinite speed, that is, speed that is instantaneous and cannot be measured. In Maxwell's formulation of Faraday's field approach, on the other hand, this information is transmitted between interacting matter with a finite speed. That is, one has to wait for the arrival of a signal before he can know that the other bit of matter is over there.

Maxwell then made the remarkable discovery that the speed of propagation of electromagnetic effects is precisely the same as the speed of light in the same medium. Once Maxwell became aware of the fact that the speed of transmission of electromagnetic forces was the same as the speed of light, he deduced that light itself is a form of electromagnetic radiation. This was important because Maxwell's discovery further generalized the unity of electricity and magnetism that was discovered by Faraday. He then incorporated all optical phenomena within one encompassing theory of electromagnetism.

It was subsequently discovered that not only were the effects of electromagnetism and optical phenomena unified in this way, but also a whole spectrum of other phenomena. It was discovered that radio propagation, infra-red radiation (for example, heat from the sun), X rays, and the colors of the visible light spectrum (from deep red to deep violet) could all be explained by a single comprehensive theory of charged matter that interacts over long distances. These different phenomena were predicted by different solutions of Maxwell's equations. Thus, the different types of corpuscles of light that Newton invoked to explain the dispersion of white light into a spectrum of different colors were explained in terms of a continuous theory involving the speed of light. According to Maxwell's theory, when white light, which is a mixture of waves of all frequencies (colors), enters a prism, each frequency wave slows down at a different speed. Thus, it takes a different amount of time for each of the colors to emerge from the prism, with the result that the observer sees a rainbow.

Maxwell's explanation is superior to Newton's explanation because it applies to phenomena other than light and it predicts the behavior of other phenomena. Another effect predicted by Maxwell's equations that could not possibly be predicted with the Newtonian model, is the phenomenon of the diffraction of light. This is the effect in which no sharp shadow of the object in the path of the light appears. Rather, the light exhibits a wave nature and "bends around corners." It should be remarked at this point, however, that with the experimental data that was available to Newton, he was not convinced that light actually did behave in this way.

A question followed from the field description of electromagnetism of Faraday and Maxwell: What is the nature of the medium through which these electromagnetic fields of force propagate? The same question also applied to the medium through which light must propagate. Maxwell was convinced that the electromagnetic waves must have some sort of medium through which to propagate just as sound waves did. This medium was referred to as the *ether*.

All the physical properties of the ether remained incapable

of measurement for a long time. Still, Maxwell and his contemporaries were convinced that it must have been present to transmit the electromagnetic radiation, for example, from a star to the earth through all of that seemingly empty space. We will see in the next chapter that an ingenious experiment by Michelson and Morley led to a negative result—that ether did not play any role in the propagation of electromagnetic radiation. This result was in agreement with Einstein's early investigations of the theory of relativity. Before Einstein formulated his theory of relativity, most physicists believed that the existence of an ether was implicit in the electromagnetic field theory. On the other hand, Maxwell's equations did not take into account the physical properties of the ether, such as its elasticity.

Those who believed in the Newtonian approach felt that all that the field equations did provide was the field of force which, in turn, must be coupled to the variables of other charged matter in order to predict the equations of motion of the latter. The coupling itself does not follow from Maxwell's equations.

It was H. A. Lorentz (1853–1928) who attempted to combine the field approach of Faraday and Maxwell with the particle approach of Newton in order to discover the equations of motion of charged bits of matter when they are subjected to the electromagnetic force field. The *Lorentz force* that was derived did provide a very accurate description of the way in which large-scale quantities of electrically charged matter would move in an electromagnetic field. On the other hand, a fundamental weakness of this theory was the absurd conclusion that when one considered the smaller and smaller charged bits of matter, the theory predicted that their energy content would correspondingly become greater and greater. In other words, the infinitesimally small atoms of matter that constitute any finite amount of material, which we observe to contain a finite amount of energy, must individually contain an indefinitely large amount of energy. As the atoms become point particles, their energy is infinite.

*16. H. A. Lorentz (1853–1928)*

Thus, a barrier appeared at this stage in the history of science that seemed to be insurmountable. So long as one insisted on incorporating the two different descriptions of matter—discrete and continuous—in the same theory, the difficulty persisted. However, it should be recalled that such an alliance was not at all the aim of Faraday's speculations. For the field concept, as it was originally introduced, took the continuous force fields as the elementary entity with which to construct a consistent description of the world.

If Faraday was correct, then it became clear that the field theory had to be understood more fully. It was fortunate that the experimental investigations of Michelson and Morley came at this very crucial time in the history of science to shed new light on the origin of the field equations themselves. These were the first of important experiments to refute the contemporary theories of matter of the late nineteenth century. These will be discussed in more detail in the following chapter.

# 6.

# Disproving classical theories

*The Michelson–Morley Experiment*

In spite of the intrinsic difficulties in the attempts to unite Faraday's field approach with Newton's atomic approach to the behavior of charged matter, Maxwell believed that this problem would be removed as soon as the properties of ether could be understood more fully. Thus scientists in the latter part of the nineteenth century inquired into ways in which they could probe the physical properties of the ether.

The experimental physicists Albert Abraham Michelson (1852–1931) and Edward Williams Morley (1838–1923) decided to carry out an experiment that was specifically aimed at detecting one of the manifestations of the ether. This was a measurement of the interference between light waves and their reflections from a mirror. (In order to make this measurement, it was necessary to know the direction of the propagating light wave, relative to the surface of the earth.) The underlying reasoning behind this experiment is as follows: Since the earth is rotating in a particular direction, there must exist some relative velocity between the earth's motion and the ether in which it is presumably embedded. Maxwell concluded that since the earth was embedded in the ether, the earth's motion would cause it

*17. Blacksmith beating iron bar while it is held in the north* (septentrio) *and south* (auster) *direction, thus making the bar a magnet.*

to drift in a direction that is parallel to the earth's surface, just as an object moving through a stream of fluid would cause a current to drag relative to it. On the other hand, there should be no ether drift perpendicular to the earth's surface. It then follows that the light waves, which are presumably being carried by the ether, should have a different speed in the direction that is parallel to the earth's surface than they would in the perpendicular direction. Consequently, by comparing the reflected waves of light from mirrors that are both equally distant from a light source, one parallel to the earth's surface and the other perpendicular, the observer should detect an interference effect because of the different speeds in the respective directions.

The reason for the interference effect is the following: Two waves start out from a common source in phase with each other, that is, the crest and trough of these waves occur at the same time. If they each travel the same distance, but with different speeds, then they would soon be out of phase with each other. Thus, when a maximum crest of one of the waves returns to the starting point in the reflected wave, the other reflected wave that returns at the same time would not have a maximum crest. By measuring the phase difference in these two reflected waves as they returned to the source, Michelson would then have been able to deduce the speed of the ether.

In the actual experiment, the results were negative! That is, Michelson and Morley did not detect any phase difference in the reflected, perpendicular light waves. This seemed peculiar, for it implied that the actual speed of the light does not depend on whether the light is propagating parallel to the earth's surface or perpendicular to it. In fact, it appeared that there is no ether. While there were those who felt confident that this negative result would be explained away, the explanation had to wait until the revolutionary concept of the theory of relativity was developed.

### Precession of the Perihelion of Mercury

If one assumes that the Newtonian law of attraction between massive bodies is correct, then a detailed mathematical analysis of the equation of motion of these bodies and their influence on each other implies that their orbits are elliptical with the other body at one of the foci (Figure 18). That is, one is assuming that the mutual gravitational force of attraction depends inversely on the square of their distance from each other. Thus, the Newtonian law implies that a planet following such an orbit, moving with a predicted cyclic motion with the sun at one focus, should return to any point on its elliptical orbit in one planetary year. This

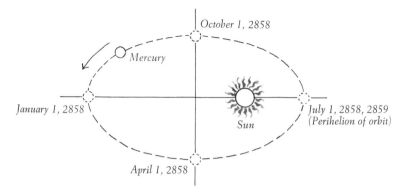

18. *A planetary orbit*

type of motion is called *periodic* because it duplicates the motion precisely, during each year.

To check the validity of Newton's law of gravitation, it should only be necessary to see if the planet will always return to the same location, relative to the position of the sun, after one year. It is easiest for the astronomer to use as comparison the particular point on the planet's orbit that corresponds to the time when the planet is closest to the sun. This particular point is called the *perihelion* of the orbit. It is convenient to focus on this point because any deviation from periodic motion in the planet's cycle would be more accentuated here, and thus easier to measure, because the orbit is curving most rapidly in this region (Figure 19). For this same reason, the most eccentric orbit (that is, the least spherical, the one that looks more like a cigar than a slightly squashed circle) is the easiest to study for the possibility of deviations.

Because of the large eccentricity of the planetary orbit of Mercury, the astronomer U. J. Leverrier (1811–1877) chose to study its motion. He checked to see if it would always return to the perihelion point in the same amount of time. It was in the middle of the nineteenth century when Leverrier observed that the motion of Mercury about the sun was indeed not periodic, that is, it did not return to the perihelion point in the same

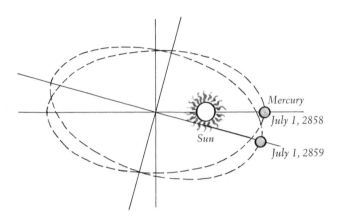

*19. Deviation from periodic motion in Mercury's orbit*

amount of time. However, it was clear, after further analysis, that one should not expect the motion to be perfectly periodic. This is because the planet is not only influenced by the sun, it is also influenced to a small extent by the existence of the other planets in the solar system. The latter effect on Mercury's orbit is indeed small in comparison with the influence of the sun; yet, it is large enough to produce a perturbation (disturbance) that would lead to a small shift in the time it takes for the planet to return to the perehelion point after each cycle has been completed.

The deviation of Mercury's orbit from a perfect ellipse due to gravitational attraction of this planet to the other planets of the solar system was calculated very accurately. It was found that while these perturbations accounted for the largest part of the observed shift, there was still about forty-one seconds of arc per century that could not be accounted for. It implied that there was some unknown force that was contributing to the deviation from the period of a planetary orbit. This anomalous (irregular) behavior was published by Leverrier in a French astronomical journal in 1859.

Leverrier's experiment was repeated, using more accurate experimental methods, at various times in the late nineteenth century and through the first half of the twentieth century. His finding was repeatedly confirmed and so there remained experimental evidence that a gravitational effect, however small, was present that could not be accounted for by Newton's universal theory of gravitation.

The negative result of the Michelson–Morley experiment on the ether drift and the anomalous result of the measured precession of the perehelion of the planet Mercury could not be explained in terms of classical theories. Indeed these results were not understood until Albert Einstein introduced the special and general theories of relativity in the early part of the twentieth century.

### *Blackbody Radiation*

Consider a box whose walls are maintained at a constant temperature. Suppose that a small window is built into the side of the box and there are removable filters for different wavelengths of radiation, that can be inserted. (These are filters that only allow particular frequencies of radiation to pass through.) An apparatus is now set up to measure the intensity of the radiation that emerges through the filters, as they are each placed in the window opening. The intensity of the emitted radiation is seen to change as a function of the frequencies that correspond to the different filters. If one should now plot the intensity of emitted radiation as a function of the frequency, then the curves obtained have the typical behavior as is shown in Figure 20. The different curves correspond to different wall temperatures.

In addition to repeating this experiment with different wall temperatures, the measurements were also carried out with boxes that were made of different materials. The measurements of boxes made of different materials led to a very perplexing result. It

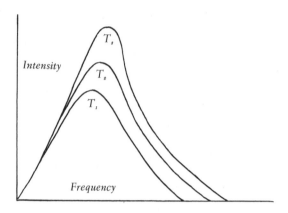

20. *Three typical blackbody radiation depend-
encies on wavelength, for successively greater
wall temperatures, $T_1$, $T_2$, and $T_3$*

was found that the curves for the spectral distribution are independent of the type of metal that constitutes the walls of the
container. This was indeed a puzzle to scientists around the year
1900, since at that time it was believed that the motion of the
charged particles of the atoms in the wall should be the source
of the electro-magnetic radiation inside of the box. But different
metals are made from different kinds of atoms and the charged
particles that constitute these atoms correspondingly move about
in different ways. It then would follow that the radiation in the
box, if it indeed represented the charged matter in its walls,
should have a different spectral distribution for different material
walls.

When it was observed that the spectral distribution curves
are independent of the material making up the walls of the box,
then the conclusion reached was that this observation involved
two independent systems. One system concerns the walls of the
box, and the other, an independent gas of radiation inside of the
box. In maintaining the walls at a constant temperature, a bal-
ance was reached between the energy transferred from the radi-

ation gas to the walls and from the walls to the gas. Such a system is said to be in thermodynamic equilibrium. When radiation is in such a state of equilibrium, it is called *blackbody radiation*.

Although unexpected, these experiments led to an inescapable conclusion about the existence of an independent gas of radiation. But the theory that explained the actual shapes of the spectral distribution curves was even more surprising. To explain these curves with a theoretical model, Max Planck (1858–1947) had to assume that the radiation gas was a collection of electromagnetic vibrations, with each one having energy that was linearly (directly) proportional to its frequency. Since the frequencies for the gas in the box had discrete numerical values, it then followed that the energy of the radiant gas must also have discrete values. This was indeed a blockbuster! For was not energy, by its very definition from the earlier studies in the mechanics of Newton and Galileo, supposed to be a continuous entity, under all conditions? What precisely did it mean to assert that energy can occur in discrete bundles—or *quanta*? (These particles Planck called quanta from Latin, meaning "how much?") In this way, Planck's innovation in explaining the unexpected experimental results of the spectral distribution of blackbody radiation led to another revolution in twentieth-century physics—the *quantum theory*.

### The Photoelectric Effect

The linear, direct, relationship between the energy of radiation and its frequency was also observed in experiments with the photoelectric effect. In this study, one shines light of a fixed frequency onto a piece of metal. The metal is, in turn, connected to an electric circuit that contains a device that can measure electrical current (Figure 21). Thus, when the light of a given color is focused on the metal, an electrical current is produced and its magnitude is measured. When the experiment is repeated

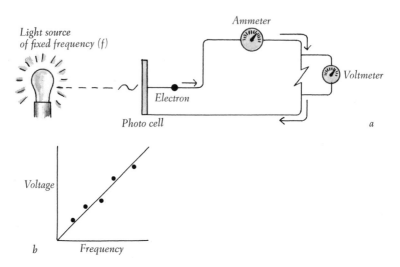

21. *Photoelectric effect.* (a) *Light of a fixed frequency hits a photo cell, causing an electric current to flow in the circuit shown.* (b) *Linear dependence of the voltmeter measurement on the frequency of the impinging light.*

using a different color light, that is, a different frequency, the generated current is seen to change. The current can be described in terms of voltage and it is the voltage that measures the energy which in turn is produced by the light beam. Finally, by studying the voltage so-produced, as a function of the frequency of the impinging light (light striking the surface of the metal), Einstein observed that the energy in the light wave must be linearly proportional to its frequency. This conclusion was in perfect agreement with Planck's earlier explanation of blackbody radiation.

These peculiar results of experiments on blackbody radiation and the photoelectric effect could then only be explained by reinstituting the earlier notion of Newton that light was indeed corpuscular in nature and further, that each of the corpuscles of light, which were later labeled *photons,* have a discrete energy that is linearly proportional to frequency. Since, it was argued, this relationship between energy and frequency is a general one, it follows that the proportion between these two quantities must necessarily be a universal constant. That is, once it is determined in one experiment, the constant is the same for all other experiments of the same phenomena. This particular constant is called *Planck's constant* and it is expressed in units of energy multiplied by time. These also happen to be the units of angular momentum. The actual series of measurements on the spectral distribution of blackbody radiation and the photoelectric effect confirmed the universal nature of Planck's constant.

While it seemed peculiar that the energy of a photon is linearly proportional to its frequency—energy in classical electromagnetic radiation depends on the square of the frequency—it became more peculiar that further experiments indicated that matter itself appeared to have this same nature. That is, future experiments implied that the elementary particles of matter that have inertial mass also have intrinsic energy that is proportional to a frequency. Thus, waves of light involved discrete particle-like features (discrete energy and momenta) and discrete bits of matter also had wave-like features (wavelength and frequency).

One of the first bits of matter to be discovered was the electron. In the latter part of the nineteenth century, J. J. Thomson (1856–1940) found that the *electron* is the smallest bit of matter to carry electrical current. However, at this stage, the classical approach to matter was not being seriously challenged. The electron was assumed then to be a small particle of matter, with a fixed electric charge and a fixed mass. The electron was assumed to obey all of the laws of mechanics that were dictated by the Newtonian approach. Thus, the momentum and energy of the particle were assumed to be continuous. Of course, as we have indicated earlier, the Maxwell field theory of electromagnetism was troublesome, for it implied that the intrinsic energy of the electron must be infinitely large. Nevertheless, many of the contemporary scientists of Thomson's day felt that

of the outer reflected wave exactly coincides with a maximum of the inner reflected wave, then the two waves would strengthen each other and then proceed together as one wave with much greater intensity than the previous, individual waves. We see, then, that if we could continuously change the diameter of the inner ring, a continuous distribution of constructively and destructively interfering waves would be produced.

The atomic nature of solid matter was discovered in the early part of the twentieth century in experiments in which electromagnetic radiation was reflected from the surface of a solid crystal. If the wavelength of this radiation was about the same as the average distance between the atoms in the solid, then one would expect that the reflected radiation from the inner layers of the crystal would interfere with the waves that were reflected from the surface of the crystal. One should expect the interference to be successively constructive and destructive, corresponding to the successive crest and trough in the intensity of all of the reflected radiation.

By the end of the nineteenth century, Faraday and Maxwell had established the wave nature of electromagnetism. In 1895, W. K. Roentgen (1845–1923) discovered X rays. X rays are electromagnetic radiation of very small wavelength emitted from the inner electrons of an atom. In the early part of the twentieth century, a father and son team, W. H. Bragg (1862–1942) and W. L. Bragg (1890–    ) observed the atomic nature of crystals by reflecting X rays from their surfaces and from the inner layers of the atoms, and studying their interference. By measuring the distances between the spots of maximum and minimum intensity in the reflected X rays, it was possible to deduce the locations of the various atomic reflecting planes in the crystals. Thus, Bragg and Bragg utilized the wave nature of electromagnetic radiation to measure the locations of the atoms that make up the solid state of matter, thereby providing added proof of the atomic nature of matter.

22. *Wilhelm Conrad Roentgen (1845–1923)*

### Electron Scattering from Solid Matter

In 1927, C. J. Davisson (1881–1958) and L. H. Germer (1896–    ), and independently, G. P. Thomson (1892–    ) (the son of J. J. Thomson) repeated the Bragg-type experiment, except that instead of X rays, they scattered a beam of electrons from a crystal. Their results implied that these particles scattered in such a way that they bunched up and then thinned out successively on the photographic plate. They did so in exact corre-

23. *One of the first photographs sent by Roentgen to demonstrate his discovery. Shown here are a set of metal weights enclosed in a wooden box.*

spondence with the way in which the X rays had successively appeared as destructively and constructively interfering waves. Thus, the electron, which was previously considered to be a small piece of solid matter like a stone, was now seen to exhibit the properties of waves. That is, it was seen that under these particular experimental conditions of scattering, the electron did not exhibit mechanical properties (momentum, mass, discreteness, energy). Rather, it exhibited wave properties (wavelength, frequency, continuity). This was indeed peculiar, for how could one think of the elementary particle as a discrete thing and as a continuous thing at the same time?

This momentous experimental result on electron scattering, while puzzling to most physicists, was expected by L. de Broglie (1892–    ). In a brilliant doctoral thesis, de Broglie showed, in 1924, how the momentum and energy of an elementary particle of matter could relate to both wavelength and frequency for the same particle. It was a conclusion of de Broglie's analysis that the energy of a particle and its frequency are linearly (directly) proportional. This was also the case for the quantum of the electromagnetic radiation field in the analysis of Planck and Einstein on blackbody radiation. It was the desire for a unified concept that led de Broglie to postulate that the constant of proportionality between the energy of the particle and its frequency was the same as Planck's constant. Imagine the excitement and intellectual fulfillment that must have ensued when the experimental studies of electron diffraction from crystalline solids confirmed de Broglie's hypothesis and determined that the ratio for the electron was indeed Planck's constant!

In summary, we see that the results of the early experiments in the nineteenth century, which confirmed the continuous field (or "wave") nature of electromagnetic forces, appeared to be refuted in experiments during the early part of the twentieth century. At this time it was implied that electromagnetic forces should be described at the fundamental level as discrete bits of electromagnetic energy, called photons. On the other hand, the "atomic" nature of matter, which seemed to have been established in the nineteenth century from the following experiments: (1) Dalton's study of the partial pressures of gases, (2) Boltzmann's statistical analysis of the thermodynamic properties of the gas in terms of an assembly of atoms, (3) J. J. Thomson's study of electron beams, and (4) Bragg and Bragg's study of X ray scattering from matter in the solid state, appeared to be refuted in the early twentieth century by the analysis of electron diffraction

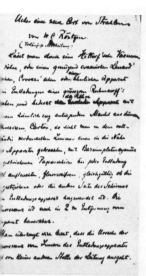

24. *From Roentgen's report announcing his discovery of X rays. The page on the left was written by Roentgen to his friend von Hippel reporting his new discovery.*

from crystalline solids. This analysis implied that, under the proper experimental conditions, these bits of matter should indeed have the continuous field description of waves.

To someone concerned with the essence of matter, this state of affairs must indeed have been frustrating. It implied that matter has a schizophrenic nature. Under some circumstances it appears as a continuous entity, and under other conditions it appeared as an assembly of discrete bits. To accept both of these approaches as a fundamental description of matter seemed to be contradictory. The twentieth-century scientist was then confronted with this question: Is the essence of matter discrete, or is it continuous, or should we be satisfied to accept both descriptions at the same time? The second question was: Are electromagnetic radiation and inertial matter the same type of stuff? After all, the earlier descriptions made them appear to be different. The electromagnetic fields of Faraday and Maxwell related to a continuous distribution of force that represents the influence that charged matter would have on other charged matter. That is, the force field is an effect, or manifestation, of charged matter. It is not the matter itself! The early twentieth-century scientist might then have stated that both the matter distribution and the force field that represents its influence have the same description!

### The Compton Effect

Important experimental studies of the nature of matter, involving a direct coupling between the electromagnetic field and inertial charged matter, were undertaken. One such experiment was carried out by A. H. Compton (1892–1962). If one should scatter electromagnetic radiation of a fixed frequency from individual charged particles (say the electrons in a gas), then the radiation scatters in a direction that can be correlated in a particular way with a change in its frequency. The frequency shift, called the *Compton effect,* depends on Planck's constant. Planck's constant appeared again in an experimental result that was accurately predicted from a theory which uses the de Broglie relation for both the scattering photon and the target electron. The correct prediction followed from considerations of the conservation of energy and momentum.

Of course, the original model that described this process in terms of only one electron and one photon is not logically consistent. For if the frequency of the photon changed in the scattering process, then the "wholeness" of the original quantum of energy is destroyed. It was assumed from the outset that the photon is indivisible! The logical consistency can be restored only if one assumes that the original photon is annihilated and a new photon, with less energy, is created in the collision process. We

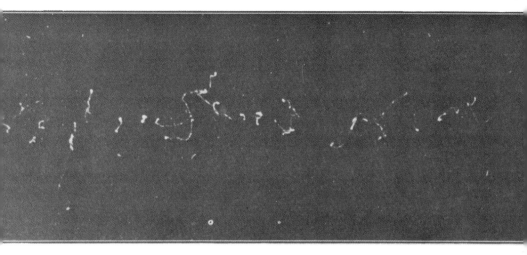

25. *Observation of the Compton effect from the ionization produced by X rays*

see that the logical consistency of the original model, that was used to describe the Compton effect, must be altered to a new model in which things are created and annihilated in collisions. However, another difficulty arises with this new model: What is the physical mechanism that predicts that when electromagnetic fields interact with the elementary inertial bits of matter, the former entities will be created and annihilated in different states of frequency? That is, energy conservation demands that because the struck electron has picked up kinetic energy, the frequency of the outgoing photon must be less than the frequency of the photon before collision. However, there does not seem to be any physical mechanism in this description that would actually destroy one photon and create another one!

### Atomic Spectra

If one should observe the dispersion of light that is produced by igniting hydrogen gas in a gaseous discharge tube and passing the glow through a triangular glass prism, it would be noted that the spectrum is not continuously distributed among the different colors that constitute white light. Instead, a set of very sharp lines would be observed. These are very bright lines of light in different portions of the spectrum. The intensities of the lines diminish very rapidly into darkness with definite displacements between them. Repetition of this type of observation, using different voltages for the gaseous discharges, and looking at different types of gases, leads to different numbers of lines at different locations on the frequency spectrum.

In the early part of the twentieth century, the spectra of different gases were studied in order to deduce further information about their constituent atoms. These studies also had some other uses. For example, by comparing the observed line spectrum of a known gas in the laboratory with the spectrum that might be observed in light that is emitted by a star, it became possible for the astrophysicists to identify the atomic composition of stars. The initial aims were to systematize the observed lines into groups. For a long period, no theory explained these empirical groups. However, with the appearance of the photon and the electron, a model of the atom was proposed by Niels Bohr (1885–1962) to explain the data. In his model, it is assumed that negatively charged electrons circulate about the central, positively charged nucleus, just as the planets of our solar system circulate about the nucleus sun. The difference in this analogy, of course, is that the force of attraction in the solar system is gravitational, while that in the atom is electrical.

A major difficulty that arises in such a model of the atom is that according to Maxwell's description of electromagnetism, accelerating charged matter must radiate energy away. The term, acceleration, refers to a change in the state of constant velocity of the particle. While the magnitude of the velocity of an orbiting electron would not change, the direction of motion of the velocity would be constantly changing. Thus a particle that is orbiting about a nucleus would be in a constant state of acceleration. It would have centrifugal acceleration. It would then follow, as a consequence of this acceleration, that the electron would be constantly losing energy and therefore it could not, according to the classical theory, stay in a fixed orbit, as postulated by Bohr.

Nevertheless, the experiments of E. Rutherford (1871–1937), in which the nuclei of helium atoms, called *alpha particles,* were scattered from atoms, gave credence to Bohr's model. The model viewed the atom as a collection of (very light) orbiting electrons which are widely separated from the (much heavier) parent nucleus.

Imagine a fat spider suspended by its web at the center of a very large room—say the rotunda of the Capitol Building in Washington, D.C. Now imagine that a flea is walking about on the ceiling, walls, and floor of the room. The relative sizes of the spider, flea, and the large room are roughly proportional, respectively, to those of the atomic nucleus, the daughter electrons, and their average mutual separation in the structure of the

*26. Apparatus used by Michelson and Morley in their experiments on light.*

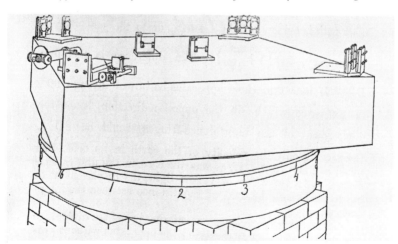

hydrogen atom. This is in accordance with the analysis of Rutherford's alpha particle scattering experiments.

Whether or not it agreed with the predictions of the classical
theory, the Bohr model did predict properties of the atom that
agreed with these experimental facts! Thus Bohr's model appeared to be valid, even though it was not possible to understand it according to the thinking of that day.

In using his model for the atom, Bohr was forced to assume
that some nonclassical concept must be in operation. He had to
adopt the nonclassical idea that the rotational motion of the
electron, which is bound to the atomic nucleus, is restricted only
to discrete spatial orbits. This is in contrast with the potential
continuum of orbits that is predicted by the classical theory of
electrodynamics. With the assumption that the orbit of the bound
electron is fixed, the electron would not be subject to the classical
energy loss. This is because such an assumption would forbid the
electron to move out of its orbit smoothly with the continual
emission of radiation. If the electron should change its orbit to
another orbit, it would now have to do so in a discrete fashion.
When such a process of energy degradation in an atom occurs,
Bohr concluded, the energy loss would be absorbed by electromagnetic radiation. In other words, a photon would be ejected
into free space. According to the quantum postulate of Planck
and Einstein, the energy of the ejected photon must then be equal
to a fixed frequency, multiplied by Planck's constant. In this way,
Bohr related the frequencies of the observed spectral lines from
radiation emitted by excited atoms to the discrete "energy-level
structure" of the observed atoms. The close agreement between
the predictions of the Bohr model and the known spectral data
was one further triumph for the discrete, quantum approach to
a theory of matter.

The implication of this agreement was that some truth must
lie in the Bohr model of the atom. Still, the model was not yet
in a satisfactory state. Many questions were still unanswered.

27. *Neils Bohr (1885–1962)*

First, while the theory did indeed refute the classical Maxwell field theory in its application to charged matter in motion, it did not yet provide a complete model to replace it. For example, where was the description for the physical mechanism that must have been responsible for the orbital electron to "jump" from one discrete level of motion to another? Exactly when did the jump occur in a given observation of a spectral line, and what was its duration? What physical mechanism describes how the photon can be created from a vacuum when an orbital transition occurs in the atoms?

Bohr's model was not yet capable of answering these questions. Yet, it refuted the predictions made by the classical theories and agreed with the atomic experiments. These experiments, the analysis of atomic spectra, the Compton effect, the photoelectric effect, electron diffraction studies, and blackbody radiation, and a host of other atomic experiments, led to a new scientific revolution. This revolution resulted in the quantum theory of measurement. In the next two chapters, we will discuss the ideas that underlie the two major revolutions that have occurred in this period—the *theory of relativity* and the *quantum theory of measurement*.

# 7

# Theory of relativity

In retrospect, we see that in the development of theories of matter from the late nineteenth century through the early part of the twentieth century, two aspects of the existing explanations of natural phenomena were in need of revision. One aspect had to do with the failure to correctly predict all of the known experimental facts, for example, the Michelson–Morley experiment (p. 54) and the precession of the perihelion of the planet Mercury (p. 57). The other aspect had to do with the logical structure of the classical theories. We have already discussed the seeming incompatibility between experimental facts and predictions of the classical theories. Let us now briefly discuss what is meant by "logical incompatibility" in a physical theory.

It is the purpose of theoretical science to hypothesize and then to elaborate on these hypotheses in order to predict the facts of nature. The purpose of experimental science is to observe and measure all of the manifestations of nature. If a particular theoretical hypothesis leads to predictions that totally agree with the experimental facts that are at hand (to within the limits of experimental accuracy), then the claim can be made that the hypothesis is true to the extent of agreeing with the known facts of nature. It is the program of theoretical science to discover the truth of nature in this way.

Such truth is always to be understood as provisional, that is, it is "true" only to the extent that it explains the known facts. If new facts are discovered that cannot follow from the original hypothesis, then the hypothesis must either be totally abandoned for a new theory, or be sufficiently adjusted (if this is possible in a logical way), so as to incorporate all of the known properties of nature in one underlying scheme. These rules are in addition to those formal procedures by which theoretical science attempts to approach truth.

In theoretical science by using the methods of deduction, one should be able to predict the observational implications of a theory from its initial axiom. These methods are the rules for approaching truth. One very important rule in this domain is the following: Given a set of assertions that constitute a particular scientific hypothesis, there must be one and only one set of conclusions that can follow. For example, if we are given a gravitational potential energy that depends on $1/R$ (where $R$ is the mutual separation between the interacting masses), the rules of prediction lead to the statement that one body must necessarily trace out an elliptical orbit about the other one, which is located at one of the foci of the ellipse. If the given rules should permit a different conclusion (for example, if it is also predicted that the orbit of one of the bodies is a circle, under the exact physical conditions that predicted the elliptical orbit), then the set of rules for this theory would be logically inconsistent. This is because the statement *A implies B and A implies C* together with the statement *B and C are mutually incompatible* cannot be consistent. That is, a given orbit is either elliptical or it is circular. It cannot be both at the same time! In the latter example it must be concluded that the set of rules that lead from the axiom $A$ to its conclusion does not obey the condition of logical consistency.

The reason for demanding logical consistency in theoretical science is clear. For it would indeed be impossible to claim that one knows something about the underlying truth of natural phenomena, if the initial physical axioms can lead to more than one conclusion in answer to the same question—even though one of these conclusions does agree with the facts. On the other hand, if a given hypothesis leads to one and only one set of conclusions, and if these conclusions agree with the facts, then one can feel fairly safe about claiming the truth of the original hypothesis.

To be logically consistent, Newton's theory leads to the requirement that "time" is absolute. This means that whoever is measuring a duration of time, whether he is at rest at one place or another, or in motion relative to a stationary observer, he will always measure the same time interval as would any other observer. On the other hand, we have seen earlier that the Maxwell field theory for electromagnetic phenomena does distinguish between the time that would be measured by one observer of

28. A stationary scientist and one who moves on a flying carpet perform similar experiments to deduce a law of nature. The principle of relativity requires that when they communicate about their results they must agree on form of this law.

*29. Gottfried G. Leibniz (1646–1716)*

these effects and another observer, who may be moving relative
to the first. Of course, it was known to Newton and to his predecessor, Galileo, that the spatial coordinates must be defined in
the laws of physics to be relative. That is, it is only meaningful
to say where something is, in relation to some other point in
space. For example, a force of mutual attraction in Newton's
gravitational theory depends on the mutual separation between
the interacting bodies. That is, the spatial location of one body
is only relative to that of another (with which it interacts). If
this were not so, then one could consider that the locations of
all bodies would be relative to the same point that is fixed in
space—an absolute center for the universe. It is recalled that the
latter hypothesis was in fact proposed by Aristotle to describe
the heavenly bodies with the earth at the center. However,

*30. Jules-Henri Poincaré (1854–1912)*

Galileo refuted this hypothesis with experimental evidence in regard to the motion of the planets and their satellites in our solar system. It was indeed a revolutionary step for Galileo to propose the relativeness of spatial coordinates in the description of the universe.

Thus, while Galileo and Newton accepted the "relativeness" of spatial coordinates, they did not question the absoluteness of time. To these scholars and their contemporaries, the "when" in a physical law must be the same for all observers. The absoluteness of the time coordinate is intrinsic in the Newtonian atomistic theory of matter.

The relativity of the spatial coordinates of the "things" in classical physics has nothing to do with Newton's own insistence that all relative, spatial coordinates must be points that are contained within a single, absolute frame of reference. The relativeness of locations of observed things was only a convenient way for Newton to locate these things in his absolute and fixed space of the universe. In other words Newton believed that there were fixed coordinates, and he located other things by saying where they were located relative to the fixed points. It is in regard to the latter opinion that historical debate ensued between a follower of Newton's view and that of Leibniz. The latter scholar insisted that an absolute space–time coordinate system could not possibly play any role in the actual description of nature, and he therefore concluded that it would not be meaningful to introduce absolute coordinates in the first place. Leibniz's view of space and time was in terms of a purely relative system of numbers. It was this approach of Leibniz that was fully exploited in Einstein's construction of the theory of relativity.

However, the field theory of Faraday and Maxwell did require one to relate the "when" as well as the "where" of one observer to another, if they were to agree about the fundamental form of the equations of the field theory as a law of nature. Thus time, as well as space, became meaningful only in the relative sense in the theory of electrodynamics and electromagnetism.

It was Albert Einstein (1879–1955) who first recognized that the electromagnetic equations required that to each moving charged particle there must correspond a "private" set of space and time coordinates. It was well known, when Einstein started his investigations, that the relativeness of the spatial coordinates in physical laws implied the following mathematical property of the space coordinates: The translation of one of the space coordinates (for example, length) in any observer's frame into the language of space coordinates of any observer in another place leads to a linear sum of all three spatial coordinates (length, width, height) in the latter observer's language. A linear sum is any sum that can be added in any multiples, as long as it is not cubed or squared. A set of three such coordinates is called a vector. Alternatively, a vector can be defined as something that has magnitude and direction. For example, when the weather report says that a thirty mile per hour wind is blowing in the northeasterly direction, a vector is referred to.

Thus, if we not only regard space as relative, but we also regard time as relative, it follows that the transformation of any one of the four space and time coordinates of one observer to the language of any other observer would generally be expressed as the linear sum of all four of the coordinates (length, width, height, time) in the second observer's space–time coordinate system. With this view, we see that it is only meaningful for one observer to relate his experience to another in terms of space and time, together. With such an approach, the world must then be described in terms of four-dimensional points. Einstein called these *events* or *world points*.

Under the special circumstances in which two different observers are at rest with respect to each other, the relativeness of their respective coordinate systems reduces to a relation between their spatial coordinates alone. On the other hand, if any two observers are moving relative to each other, then a generalization of Einstein's conclusions to any law of dynamics implies that in making the transformation of coordinates from one observer to another, all four space and time coordinates must be treated together. If this relative motion corresponds to a constant speed in a straight line, then the different frames of reference are called inertial and the study of these frames of reference comprise the theory of *special relativity*. According to this theory, the laws of nature must be unchanged in form when undergoing the transformations that relate to any two inertial frames of reference. The latter "translations" of the language of one inertial observer to that of another are called *Lorentz transformations*. For it was Lorentz who first discovered the mathematical form of the transformations of space and time coordinates which would ensure that the form of Maxwell's equations would be the same in all inertial frames of reference.

As indicated in the earlier discussion of the structure of the Maxwell field equations, a constant speed must be introduced in order to maintain the invariance of the laws of nature required by Lorentz transformations. That is, to convert a time dimension to a combination of time and space dimensions, the units of time must be expressed in terms of the units of length. This is done by introducing a universal speed which multiplies all time coordinates to give them the dimension of length. Maxwell discovered that this speed is precisely the speed of light in a vacuum. It represents, more generally, the speed of propagation of interaction between the mutually combined parts of a physical system. The theory of special relativity is structured in such a way that when the relative speed between inertial frames of reference becomes small compared with the speed of light, then the predictions of the theory reduce to the standard predictions of the classical theories. In this limit, the Lorentz transformations of relativity theory reduce to the so-called *Galilean transformations* of classical mechanics. An example of the latter is the transformation between the coordinate $x'_G$ in a fixed frame and the coordinate $x$ in a frame that is moving at the relative speed v. The relation is $x'_G = x + vt$. The corresponding Lorentz transformation from which this Galilean transformation is derived is:

$$x'_L = (x + vt)/\left(1 - \frac{v^2}{c^2}\right)^{1/2}.$$

We see here that if the ratio $v/c$ (where $v$ is the relative speed and $c$ is the universal speed) is sufficiently small compared to unity, to allow us to neglect it, the two transformations, $x'_G$ and $x'_L$ are the same.

*Cours professé à la Faculté des Sciences.*
*par M. H. Poincaré.*
*1886-1887.*

# Cinématique.

## 1ère Leçon.

1er mai 90

La cinématique est l'étude des mouvements indépendamment des causes qui les produisent, ou plus exactement, c'est l'étude de tous les mouvements possibles.

A côté de la notion d'espace qui fait l'objet de la Géométrie. La Cinématique introduit en outre la notion du temps.

Toute figure mobile peut être regardée comme un système de points mobiles; il est alors naturel de commencer par l'étude du mouvement du point mobile isolé.

### Mouvement rectiligne uniforme.

On dit qu'un point mobile décrit un mouvement rectiligne uniforme, quand il se déplace sur une droite, de telle sorte que les espaces parcourus sont proportionnels aux temps employés à les parcourir.

Soit $OX$ la droite sur lequel se déplace le point $M$.

Soit $x$ la distance du point $M$ au point fixe $O$.

Les accroissements de $x$ sont proportionnels aux accroissements du temps $t$, ce qui montre que $x$ est une fonction linéaire de $t$:

$$x = at + b$$

Dans ce mouvement on appelle <u>vitesse</u> l'espace parcouru dans l'unité de temps

Au bout du temps $t$ l'espace parcouru est :

$$x = at + b$$

Au bout du temps $(t+1)$ l'espace parcouru est :

$$x_1 = a(t+1)+b$$

La longueur parcourue dans l'unité de temps est donc la différence :

$$x_1 - x = a$$

Imp. C. Lacube & ses Fils, 11 rue Madame, Paris.

Poincaré N° 1.

---

31. *First lesson from one of Poincaré's books teaching mathematics*

The Lorentz transformation between the time coordinates is similar to the preceding transformations between the spatial coordinates of two different inertial observers. Again, if we neglect the ratio $v/c$ because of its smallness compared with unity, the transformation becomes a Galilean transformation of the time coordinate. That is, time is the same in all inertial frames in the nonrelativistic theory. However, it should be emphasized that from the point of view of relativity theory, the nonrelativistic theory is only a mathematical approximation in which it appears that the time is absolute. Such an appearance is, of course, strictly illusory, so long as one adheres to the relativity theory, since it follows from an exact form in which the time coordinate is relative to the inertial frame.

The reader should be cautious at this point to resist the temptation of thinking of time, as it is experienced, in the same way as one normally thinks of space, as it is experienced. At this point one might imagine that if space and time are to be treated alike, then we should be able to travel in time just as we do in space! For example we should be able to take a trip into the past or into the future. To accept this possibility, one would have to give an entirely different interpretation of the time (as well as the space) coordinates as compared with their interpretation in the relativity theory. Indeed, the revolutionary change that was proposed by Einstein replaced the absoluteness of space and time coordinates with their relativity. This means that an observer merely uses the space and time coordinates as a convenient language with which to express his experience. Any different observer who might be somewhere else, at a different time, and in motion relative to the first observer, would express his experience in his own language, that is, in terms of his own space and time coordinates. The assertion of the theory of relativity is that when one translates from the language of one observer to that of another, then if the laws of nature deduced from their respective experiences are to be valid laws, they should agree about

Einstein agreed with Poincaré's first view. However, he argued that the type of geometry that should be used to define the relations between points in space and time must not be arbitrary. Einstein did agree that in the domain of dimensions that are used to make the measurements in our immediate surroundings the geometry should correspond to the Euclidean rules because our actual measurements make it appear to be so. To clarify the distinction between Euclidean and non-Euclidean geometry, we can make the following general statement. Euclid's geometry basically consists of a set of axioms that are absolute and seem to conform to our experience. For example, it is assumed in Euclidean geometry that the shortest distance between two points is a straight line. On the other hand, a different set of axioms, which may define a different sort of geometry, may not imply that the shortest distance between two points is a straight line. The mathematician G. F. B. Riemann (1826–1866) was the discoverer of a particular non-Euclidean geometry and Einstein was later able to show that Riemann's geometry provided a more accurate picture of the universe as a whole.

Thus, while Euclid's geometry led to statements about the relations between the space–time points that were in quite close agreement with observations in our immediate surroundings, Einstein showed that a more accurate description was non-Euclidean. That is to say, the Euclidean relationships served only as a mathematical approximation for something that is not actually Euclidean. One of the important results of Einstein's study was that with the Riemannian type of geometry to describe space and time, and with no other assumptions, all of the properties of gravitational forces automatically follow. Just as the Riemann geometry is a special non-Euclidean system that becomes Euclidean in the "local" domain of small (that is, non-astronomical) dimensions, so Einstein's theory led to the Newtonian theory as a good mathematical approximation in the local domain. Still, a fundamental difference between relativity and the classical

Preclarissimus liber elementorum Euclidis perspi/
cacissimi:in artem Geometrie incipit quafoeliciffime:

Punctus est cuius ps nõ est.ℂLinea est
lõgitudo sine latitudine cui⁹ quidé ex/
tremitates st duo pũcta.ℂLinea recta
é ab vno pũcto ad aliũ breuiffima exté/
sio i extremitates suas vtrũq3 eoꝝ reci
piens.ℂSupficies é q̃ lõgitudine ⁊ lati
tudiné tm̃ h3:cui⁹termi quidé sũt linee.
ℂSupficies plana é ab vna linea ad a/
liã extésio i extremitates suas recipiés
ℂAngulus planus é duarũ linearũ al/
ternus ꝑtactus:quaꝝ expãsio é sup sup/
ficié applicatioq3 nõ directa.ℂQuãdo aũt angulum ꝑtinét due
linee recte rectilineꝰ angulus noiaꝑ.ℂQñ recta linea sup recta
steterit duoq3 anguli ytrobiq3 fuerit eq̃les:eoꝝ vterq3 rectꝰerit
ℂLineaq3 linee supstãs ei cui supstat ꝑpendicularis vocaꝑ.ℂAn
gulus vo qui recto maioꝛ é obtusus diciꝑ.ℂAngulꝰyo minoꝛ re
cto acutꝰappellaꝑ.ℂTermin⁹é qd vniuscuiusq3 finis é.ℂFigura
é q̃ tmino vltermis ꝑtinét.ℂCircul⁹é figura plana vna q̃dem li/
nea ꝑteta: q̃ circũferentia noiaꝑ:in cuiꝰmedio pũctꝰé: a quoꝰoés
linee recte ad circũferétiã exeũtes sibiuiceꝫ sut equales. Et hic
quidé pũctꝰcétrũ circuli dr.ℂDiameter circuli é linea recta que
sup ciꝰcentꝛ trãsiens extremitatesq3 suas circũferétie applicans
circulũ i duo media diuidit.ℂSemicirculus é figura plana dia/
metro circuli ⁊ medietate circũferentie ꝑtenta.ℂPoꝛtio circu/
li é figura plana recta linea ⁊ parte circũferétie ꝑteta: semicircu/
lo quidé aut maioꝛ aut minoꝛ.ℂRectilinee figure sũt q̃ rectis li/
neis cõtinent quarũ quedã trilatere q̃ trib⁹rectis lineis: quedã
quadrilatere q̃ q̃tuoꝛ rectis lineis. q̃dã mltilatere que pluribus
q3 quatuoꝛ rectis lineis continenꝑ. ℂ Figurarũ trilaterarũ:alia
est triangulus bñs tria latera equalia.Alia triangulus duo bñs
eq̃lia latera.Alia triangulus triũ inequalium laterũ.Paꝝ iterũ
alia est oꝛthogoniũ:vnũ.s.rectum angulum habens.Alia é am/
bligonium aliquem obtusium angulum habens.Alia est oꝛigoni
um:in qua tres anguli sunt acuti.ℂFigurarũ auté quadrilateraꝝ
Alia est q̃dratum quod est equilaterũ atq3 rectangulũ. Alia est
tetragon⁹long⁹:q̃ est figura rectangula : sed equilatera non est.
Alia est helmuaym: que est equilatera : sed rectangula non est.

De principijs ꝑ se notis:⁊ pmo de diffini/
tionibus earandem.

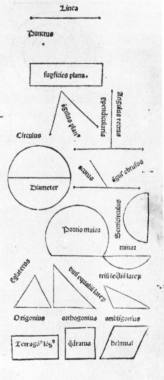

theories in regard to the use of space and time coordinates is that in the Einstein approach, the explicit relation between the space–time points must be derived by solving basic field equations, while the geometry in classical theories was explicitly specified beforehand.

It might be mentioned at this point that Poincaré's view of space and time was somewhat more liberal than that of the eighteenth-century philosopher Immanuel Kant (1724–1804). It was Kant's view that the geometry of space and time must necessarily follow the axioms of Euclidean geometry. He took this view because he insisted that our "common sense" requires it. Of course, by the term, "common sense," Kant was referring to our opinions of the world as a result of our experience—through our senses. For example, it seemed intuitively obvious to Kant that given a straight line and a point outside of it, there could be one and only one other straight line that would pass through this point and never cross the first straight line. It is true that this was one of the axioms of Euclidean geometry and that one cannot prove the truth or falsity of an axiom because it is, by definition, something that is given. Yet, Kant felt that the truth of this axiom was dictated by "common sense." It should be remarked, however, that Euclid's was the only geometry that existed in Kant's day. It is hard to say, in retrospect, whether Kant might have changed his mind had other geometries been discovered that did not contain Euclid's axiom about parallel lines. Nevertheless, as he stated it, Kant's approach required Euclidean geometry to be the true one for intuitive reasons.

During the nineteenth century, non-Euclidean geometrical systems were discovered that did not, for example, contain Euclid's axiom of parallels. Thus, the way was then paved for adopting a more liberal view of space and time, such as the conventionalist approach. Poincaré did not claim with Kant that Euclidean geometry was the only one that had to do with reality because of "common sense." Rather, he accepted the possibility that any geometry might be used to describe the physical world. He argued in favor of Euclidean geometry only because it was the simplest one to handle from the mathematical point of view. Einstein followed Poincaré's lines of reasoning and agreed with his view of the use of space and time coordinates in order to facilitate a description of physical experience. (Such was the view of Faraday fifty years earlier.) However, Einstein disagreed that the true description of the world would be insensitive to the particular type of geometry to be used. Thus he was ready to begin with the theoretical development of the theory of general relativity.

Should one accept the idea that the space and time coordinates are only relative entities to be used in the language of physics, the question might then be asked: What is the "real stuff" from which the universe must be described according to this view of matter? Einstein might have answered as follows: In any description of nature there must be an "observer" and the "observed," who communicate with each other in terms of signals that are passed between them. For example, the "observer" might send out a pulse of light that would subsequently be absorbed by the "observed," a piece of crystalline matter. After absorbing the signal, the crystal would become "excited," but it would eventually "calm down" by re-emitting another signal back to the "observer" emitter. From the latter reaction of the crystal, the "observer" may deduce some of its properties (as in the Bragg experiment on X ray reflection from crystals). However, suppose that the crystal had not re-emitted the signal. Then as far as the "observer" is concerned, the crystal may as well not exist. For the "observer" to say that the crystal does exist in this case would only be idle speculation!

When one carries the approach of relativity theory to its logical extreme, it should make no difference if we had called the "observer," the "observed," and vice-versa. These names are only relative because the interaction between them is really independent of which is called "observer" and which is called "observed." It is the interaction itself—the mutual relation—that must be considered as the objective entity in any description of matter, according to the relativistic approach.

Let us exemplify this idea with the following allegory. Consider a small flea to live on a particular, small region on the back of a large, hairy dog. Let us say that (from the human time scale) the dog has a life span of about fifteen years and the flea has a life span of about three days. Now suppose that we can communicate with the flea and we ask it to tell us about the world, as he sees it. If the flea should take the Newtonian approach, then he will tell us that he lives in a warm forest of tall, soft trees. (The Newtonian approach concentrates on things rather than the relationship between things.) The ground is undulating very slowly. However, the flea would continue, every few years there is a major earthquake with the ground shaking so vigorously that I am thrown out into the blue sky, only to land in a new forest of similar, tall, soft trees and warm, undulating ground. After describing his surroundings in this way, the flea would proceed to tell you about himself—his brilliant method for finding food, his way of having fun, and a whole list of other self-attributes.

If the dog is also "Newtonian," he would complain about a severe itch on his back. He would tell you that he shakes himself quite vigorously every few minutes, but the itch always appears again on a new spot. Sometimes it takes more than three days for the itch to go away altogether, he would sigh. The dog would then go on to tell you all about his self-attributes. Thus, after we had listened to the flea's and dog's story about the physical universe, we would notice that we had heard two entirely different stories about only one world! Whose story is the more reliable one? It would be difficult indeed to decide.

The approach just discussed is the classical or Newtonian one in which the essential entities that constitute the world are

*33. Relativity and Objectivity. Whose description of the world is correct, the dog's or the flea's?*

the individual things. Their descriptions of the world are necessarily self-centered. Since the dog and the flea are different sorts of things, we would expect their descriptions of the universe to be different. Neither one of them is to be considered more reliable than the other since they both told incomplete stories.

Suppose now that the flea and the dog had taken the Einsteinian, rather than the Newtonian view. Einstein's view as opposed to Newton's view, is that the relation between things is more fundamental than things themselves. Here, it is the mutual relation that is to be considered as the basic building block from which to build a description of the universe. In this case, the flea would proceed as before, telling you all about the warm forest with tall, soft trees, the undulating ground, and the occasional earthquakes. However, he would not proceed to relay to you all of his self-attributes; instead, he would make a deep and more probing study of his world. After completing this study he would then relate to you the earth's (that is, the dog's) reaction to his presence. Thus, the sum total of his experience would be in terms of his reaction to the dog and the dog's reaction to him. None of the self-centered thoughts would occur to the flea. Similarly, the Einsteinian dog would tell you about his reaction to the flea (just as before), as well as the flea's reaction to him. After deciphering the languages of the flea and the dog, we would see that in the latter approach to a description of the universe the flea and the dog would be telling precisely the same story—a story in which it is the mutual relation that is the basic element, and where the story is consequently complete and therefore entirely objective.

Instead of the "thingness" of the Newtonian approach to a basic description of matter, it is the mutual relation that is the real stuff from which to build a description of the laws of nature, according to the relativistic theory. This is not meant in the sense that there are actual distinguishable parts that might later be coupled in one way or another, but could always be uncoupled without losing their individuality. The latter view would indeed take us back to the Newtonian approach. When we say that it is the mutual relation, or the interaction, that is the elementary stuff, we mean that the physical system described is whole; it is fundamentally *one;* it is not the sum of parts. In the metaphor about the dog and the flea, the oneness was the dog–flea—a word that is defined in a way that prohibits the removal of the hyphen. The illusion of the dog or the flea as separated entities is here only an approximation that is useful to describe the *closed system,* dog–flea, when there is sufficiently weak coupling. But it is important that with this view, no matter how weak this coupling may be, it can never be "off" altogether. For the separated parts do not exist in this view. The appearance of almost separated parts is only a manifestation of a single closed system. It is comparable to a small ripple on a pond. The ripple is not an independent thing. It is a manifestation of the entire pond.

This concept of *oneness* in a description of the world was

a revolution in thought that was proposed and discussed by philosophers during many of the preceding centuries. However, it took the great twentieth-century physicist and philosopher Albert Einstein (1879–1955), to develop this idea of relativity in the construction of a mathematical language that could make concrete, quantitative predictions about the observable properties of nature. Let us now examine the successes and failures of this development.

## The Containment of Classical Physics within Certain Limits

As indicated earlier, Einstein's first predictions from the theory of special relativity had to do with the forces that are exerted by electrically charged matter in motion. He required, from the outset, that a correspondence principle must always be present: every affirmative test of classical theory must have a corresponding proof in relativity theory. Thus, at some appropriate limit (boundary), the equations of relativity theory should reduce to the classical equations of motion. But what should be the criterion that defines "appropriate limit"? Clearly, this should refer to the situation in which it appears as though the time coordinate is independent of the observer. The interaction between the components of a system propagate at a finite speed $c$, while in the action-at-a-distance approach of classical physics the interaction propagates at an infinite speed, that is, any part of such a system in the latter approach would "know" about the insertion of another part into the system instantaneously. One part reacts to the existence of another simply by virtue of its being there in space. Thus, if we consider the universal speed $c$ to be infinite, compared with the relative speed between inertial frames in Einstein's theory (that is, if the ratio $v/c = 0$), then the relativistic equations of motion should take the form of the classical equations of motion. (Classical physics was successful, as far as it went, because it dealt with speeds that were much slower than the speed of light.)

All of the predictions of the classical mechanics of Galileo and Newton are also predicted by Einstein's "relativistic mechanics." Yet classical mechanics and relativistic mechanics are entirely different, both in concept and in the exact forms of their equations. The classical expression of mechanics is only an approximation of the relativistic expression of mechanics. The reason that classical mechanics was acceptable for such a long period of time was, in this view, that all of its predictions had to do with experimental situations in which the relative velocities of interacting things were very much smaller than the universal speed $c$ (for example, the velocity of the earth relative to the sun or the velocity of a block that is sliding down an inclined plane).

### Tests of the Theory of Special Relativity

Let us now examine the predictions of the *theory of special relativity* in those cases where the relative speeds between inertial frames are not small compared with the universal speed $c$. The first obvious example has to do with the propagation of light, since its speed in a vacuum is just equal to the universal speed itself. The first prediction, that would not be acceptable within the classical approach, is that since $c$ is a universal constant, the actual speed of light is independent of the frame of reference in which it is described. In other words, the speed of light would be the same in any situation. Thus, for example, the net speed of emitted light would be the same if the source of light was stationary or in relative motion. To compare this with the classical case, think of a man standing on the ground, watching a ball that is thrown from the window of a moving train. The velocity of this object would be the sum of velocities of the ball, relative to the floor of the train, and the velocity of the train, relative to the ground. That is, the velocity of the source (the train) must be added to the velocity of the ball to obtain its resultant velocity relative to the ground. If this ball was, instead, a beam of light and if we were invoking relativistic, rather than classical mechanics, then the velocity of the source would play no role. The velocity of the light wave relative to the ground would be the universal speed, whether it was emitted on a moving train or on the ground.

The preceding conclusion does not necessarily imply that "light" is an independent entity. The conclusion merely relates to the speed with which one part of an interacting system communicates with another part. This result states that the speed of "communication" has a maximum value and that this speed is independent of the particular frame of reference in which the interaction is measured. It implies that the speed of light parallel to the rotating earth's surface is the same as its speed perpendicular to the surface. This conclusion is then in agreement with the negative result of the Michelson–Morley experiment in which the speed of the ether drift was found to be zero. The ether does not increase the speed of light. The implications of these experimental results, together with the theory of special relativity, is that there is actually no need to describe the propagation of electromagnetic fields in an ether; that is, the existence or nonexistence of an ether to carry fields of force (as originally assumed by Maxwell) plays no role in the physical description of electromagnetism. Thus we see that the theory of special relativity was the first analysis that gave some theoretical understanding of the previously peculiar-looking results of the Michelson–Morley experiment. But such understanding was not achieved by patching up the classical theory. It came about, rather, by overthrowing a part of the foundation of the classical theory. It required a new approach, a passive approach to space and time coordinates, as the elements of the language of some observer. This is in contrast with the active role that is played by the space–time coordinates in the classical theory where all observers must be cast into an existing real and absolute frame of reference to describe nature.

Another interesting prediction of the theory of special relativity is the shift in frequency of an electromagnetic wave (say a light wave) that is moving relative to a fixed observer. The observations of these phenomena (for example, the aberration of light—the nonfocusing of light of different colors as they pass through spherical lenses, and the Doppler shift—the variation of frequency noticed when the source of the light and the observer are moving relative to each other) are numerically in agreement with the predictions of the theory of relativity. They are not in agreement with the classical theory. In both the classical theory and the theory of relativity some value for the Doppler shift is predicted. However, the values predicted in each case are different. Scientists have proven that the value predicted by the theory of relativity was the correct one. For example, when a moving train passes a fixed observer, he will hear a sound with continuously changing frequency (from the high to the low notes). The calculation of this frequency shift in classical theory depends on the application of transformations from a moving object (the train) to a fixed object (the observer)—with a time that is the same for both. In relativity theory, the time scales are different for the relatively moving frames of reference. Thus, the prediction of the Doppler shift in the classical theory (because of the relativity of space coordinates alone) and the prediction in relativity theory are different. An exact measurement of the Doppler shift for light waves, when compared with the different predictions of each of the theories provided a good test. Needless to say, the theory of relativity passed this test and the classical theory failed.

What about inertial matter that moves at speeds which are close to the universal speed $c$? Once again, measurements on such systems can provide a good test for the selection of either the classical theory or the relativity theory as the more valid one. A typical test involves the measurement of the distribution of energy and momentum in two massive objects that have collided with each other at very high speeds. The theoretical implications of each of these theories makes us analyze each experiment quite differently. Thus, the very good agreement of the actual data on high-energy collisions with the relativistic predictions once again proves the validity of relativistic mechanics over classical mechanics. The relativistic description of collision processes between massive particles reduces exactly to the classical description if we allow the speed of light to be considered as infinite, compared with the relative speed of the colliding systems.

Another important consequence of relativity theory that does not follow from classical mechanics is the prediction that even when a massive body is at rest and without any potential energy, there still remains a fixed quantity of energy in this body. This is called the rest energy. Analysis of relativistic mechanics predicts that the rest energy of a body is equal to its inertial mass multiplied by the square of the speed of light. An astounding consequence of this formula is that the most inconsequential-looking piece of matter, say a worn penny, contains a tremendous amount of dormant energy. For, if the mass times the speed of light squared, of a penny, could be converted into utilizable energy, then it could easily light up the city of New York for many months. The discovery of a means to tap this energy by means of nuclear fission was a major discovery of the twentieth century that not only led to a new technology, but also, to a further confirmation of the theory of relativity.

A practical means of utilizing nuclear energy by means of the nuclear fission process was discovered by E. Fermi (1901–1954). The idea is the following: When some very heavy atomic nuclei are made even heavier by their absorption of nuclear particles, they become highly unstable. Instead of simply emitting the original particle that was absorbed, the unstable nucleus will sometimes split into two or three lighter nuclei. This is called nuclear fission. The sum of the masses of the lighter nuclei is usually less than the mass of the original unstable nucleus. The difference in terms of rest energy has been converted into the measurable kinetic energy of the fissioned parts, thereby verifying Einstein's theoretical result.

When this happens to the $U^{235}$ nucleus after it is bombarded with a slowly moving neutron, the heavy nucleus not only fissions, but it also emits two neutrons in the process. Thus, from one neutron there comes two. Each of the two neutrons then bombards two other uranium nuclei; subsequently each of these fissions gives off two more neutrons each. Thus, starting with one neutron, two are produced, and then 4 more and then $2^3 = 8$ more, and in no time at all $2^N$ neutrons can be released, where N is of the order of $10^{23}$ when the piece of uranium is a laboratory specimen. Thus a chain reaction ensues and each time a fission occurs there is a release of kinetic energy. This chain reaction is thus capable of releasing huge amounts of utilizable energy.

The story is told that in the 1930s Fermi was trying to produce this chain reaction by bombarding uranium with highly energetic neutrons. He had no luck. One day, when he was running across the campus court at the University of Rome with a neutron source, a piece of uranium and a Geiger counter in his hands, he tripped and the uranium and the neutron source fell into a fish pond. Suddenly, the Geiger counter started to click away at a very rapid rate, indicating that a very great amount of energy was being released from the uranium sample. A chain reaction had been produced. Fermi immediately realized that a chain reaction could only be started by "cold" neutrons—that is, neutrons that move at a very slow speed, as compared with the high-energy "hot" neutrons with which he had been experimenting. The water in the fish pond had acted as a moderator to slow down the neutrons that were originally emitted with high energy from his radioactive source. This interesting accident, and Fermi's understanding of the results that followed, was the initiation of the twentieth century to the age of nuclear technology.

## General Relativity

The successful predictions of the theory of relativity that we have been discussing so far have had to do with the mechanics of light (for example, the photon—a quantum of light energy) or massive objects, that are in inertial frames of reference. Such frames are relatively at rest or in relative motion in a straight line with a constant speed. According to the laws of Galileo and Newton, such objects at rest, or in constant speed relative to other objects, will maintain this state of motion *forever*, unless compelled to change this state by some external force. According to the principle of relativity, all inertial frames must be equivalent: One can transform the laws of nature (in terms of the fundamental equations) from one inertial frame to any other without noticing any difference. But does the inertial frame of reference entail the most general sort of language with which to describe the laws of nature? Obviously not! In the realistic situation, where the "parts" of a system would *interact* with each other (for example, a measuring device and the "thing" that is measured), these components would not be in inertial frames because they would be exerting a force on each other by virtue of their mutual interaction. Thus, if we are to treat the real problems of physics from the point of view of a full exploitation of the principle of relativity, the most general coordinate frames would indeed not be inertial.

Of course, when two massive objects scatter off of each other, their mutual interaction takes place only during the very short period when they collide. Here, the motion of each of the bodies is uniform (constant speed in a straight line) before and after the collision. At these times it is permissible to use a formulation that agrees with special relativity theory. But so long as the interaction is "on," it is not strictly permissible to use special relativity theory in describing the motion of one of these objects relative to the other. For example, consider two billiard balls approaching each other on the billiard table. At most times they are unaware of each other's existence, and they proceed along their respective straight-line paths with constant speed. For a very short time they will interact. After the collision, they then proceed once again to move in straight lines with constant speed, but in a different direction than previously. Special relativity theory then applies at times before and at times after the collision, but not during the collision.

It should be noted at this point that in reality, the mutual interaction within a physical system is never really "off," as it was in the idealization of the previous example. Consider the Coulomb force, or the gravitational force between interacting bits of matter. These forces depend on $1/r^2$, where $r$ is the mutual separation between the interacting things. These forces are exactly zero only when the bodies are separated by an infinite distance. Nevertheless, if $r$ is sufficiently large, $1/r^2$ can be approximated by zero, without losing too much accuracy. With the replacement of the very small number, $1/r^2$, with an actual zero, special relativity theory can then be used.

In the realistic case, however, where account is taken of the fact that $1/r^2$ is not actually zero at ordinary separations, the motion is not exactly uniform and, in principle, the motion must be described with a law of nature that conforms to the *principle of general relativity*—the assertion that the laws of nature are the same in all space–time frames of reference, irrespective of their motion relative to each other, be it uniform or nonuniform. Thus we see that so long as $1/r^2$ (in this case) is not exactly zero, the mathematical predictions of special relativity theory can only be considered as an approximation for the predictions of a theory of general relativity.

The principle of general relativity, then, requires that if any two observers are in relatively noninertial frames of reference (that is, they are moving nonuniformly with respect to each other, for example, by virtue of the circular motion of one observer relative to the other), then if they should separately deduce the laws of nature from their experiments on similar types of phenomena, they should agree about the form of the laws of nature so-derived. Of course, these observers would never know if their respective deductions did indeed have the same form until they learned to translate the language of one space–time reference frame into that of the other. These translations of the "words" (space–time points) of the respective observers are continuous, coordinate transformations. They are continuous because the different reference frames are distinguishable according to their relative motion, and motion, in turn, is defined as a continuous change of the space–time coordinates with respect to each other.

It then follows from the principle of relativity that once one learns the coordinate transformations that will maintain the form of one particular law of nature, the same transformations of coordinates must apply to the expressions of all of the laws of nature. With this, the fundamental starting point for the mathematical formulation of the theory of general relativity is specified.

Thus far, we note three important features in the theoretical approach of general relativity. First, there is no special requirement at this stage to adopt a "language" of space and time coordinates based on the geometry of Euclid. Second, the transformations from the space–time coordinates of one observer to those of another in this general coordinate system involve only continuous changes. Third, if the coordinate frames that are used are not inertial, then the effects of forces are already incorporated in the language that any given observer must use to describe his environment.

Einstein was very much intrigued by the latter conclusion,

34. *Euclid greets students at the outer gate of a circle which surrounds figures who represent the mathematical disciplines: arithmetic, geometry, music, astronomy, astrology.*

that the effects of forces are included in the language used by an observer in a noninertial frame of reference. He proceeded to take advantage of this conclusion by studying the properties of freely moving objects in a *non-Euclidean* space that was to be related to noninertial, coordinate frames. Einstein's aim, at first, was to see if instead of introducing external forces *ad hoc*, he could derive the force of gravity as an implicit feature of a more general type of geometry than the one which was used by Galileo and Newton. As we have discussed earlier, the mathematician Riemann had already developed a non-Euclidean geometry which has the property of *approaching* the Euclidean case at the limit where the points in this space come very close together (the "local limit"). In this sense, Riemannian geometry is a *generalization* of the Euclidean geometry which Newton, Galileo, and Kant felt to be the only conceivable one with which to describe a physical theory.

Thus Einstein arrived at a set of equations that would tell an observer about the details of the space–time coordinate *language* that he must use (most generally) to derive the properties of any system as a function of its matter content. With this theory, if one is given a complete description of the matter content of a physical system, then it should be possible, *in principle*, to solve Einstein's field equations for the field variables that relate to the shape of the coordinate space in which the observer must describe his experience. Einstein found that to derive an apparent force from the geometrical properties of space–time alone, the shape of space–time has to be curved. This is in contrast with the "flatness" of Euclidean space.

To demonstrate the difference between "curved" and "flat" spaces, it might be instructive to consider once again Euclid's axiom of parallels. Instead of the realistic four-dimensional space–time, consider a two-dimensional surface. It is clear that if such a surface is perfectly "flat," then if we are given a straight line in this surface and one point in the surface that is not on the line, there will be one (and only one) straight line that contains that point and never crosses the original line.

On the other hand, if the space is a closed, spherical surface and if one defines "straight line" in this space as one of the circumferences of the sphere, then given one such "straight line" and a point outside of it in the spherical surface, *all* other "straight lines" through this point will cross the original "straight line." Thus, there are no parallel lines within such a geometry. The latter property of space–time, in which there are no parallel straight lines, is a feature of a "curved" Riemannian space. "Straight" relates to the path of the shortest distance between any two points.

The reader should note that there is not a complete analogy between a Riemannian curved space and the surface of a sphere. For example, if one looks at the surface of a sphere from the inside, it is concave (positive curvature); if one looks at it from the outside, it is convex (negative curvature). The Riemannian space has a curvature that is either concave or convex in whatever direction one may look. With this difference, it should be further noted that the Riemannian space is not one that can actually be seen in terms of our senses, as we see the surface of a sphere. However, the physicist does not claim to "see" the points in a Riemannian space. He only claims to *use them* in a general description of the variables of a field theory.

Thus, it was found in Einstein's investigation that the description of a freely moving object in a Riemannian space–time is equivalent to describing the same body in a Euclidean frame of reference with the addition of an external force. The force that Einstein derived from the geometrical properties of space–time that is curved is, then, not a force in the conventional sense, that is, it is not an externally imposed force. It is rather a property of space–time itself that makes it appear that without the imposition of any external force the freely moving body would still accelerate. At the local limit, the external "apparent force" takes on the exact form of Newton's gravitational force between massive bodies.

Since the "local limit" refers to those regions of space–time where Newton's law was shown to be valid, according to the experiments at hand, Einstein's equations must be valid in this same region. It is only that the general theory of relativity, according to Einstein's view, would interpret these local observations in a different way. For example, the force of gravity between the earth and the sun is not related here to an external force that each body exerts on the other at a distance. Einstein would rather say that the earth and the sun are moving freely in a curved space–time. While the predictions of the two theories agree in the local limit (that is, in the immediate area of a particular point where all of the successes of the Newtonian theory are exhibited), the two theories are conceptually and mathematically entirely different. The reason the two agree in the immediate area of a point is because space always appears to be flat over such relatively short distances. (Even though we know the earth is round, it looks flat to us, even when we are standing in the middle of a large area.)

Conceptual differences in the two theories are, in the most general sense, that Einstein's theory is a field theory in which interactions are propagated at a finite speed and continuous entities are the essential elements while Newton's theory is a particle theory in which there is instantaneous action-at-a-distance and where discrete entities are the essential elements. The former approach is based on the relativity of space and time, while the latter is based on their interpretation as absolute and given. Thus, to choose between them, the philosopher will ask which of these two theories is more satisfactory from the point of view of logic. The physicist will ask which of these theories is more satisfactory from the point of view of predictions of actual physical phenomena. We have already seen that there is nothing that the Newtonian theory predicts that is not also predicted by Einstein's theory. It then remains to be shown that the latter theory predicts physical observations that are not at all predicted by the classical theory, if the physicist is to favor Einstein's approach.

Indeed, there were three predictions made by the theory of general relativity that agree with experimental observations and are not at all predicted by Newton's theory of gravitation. The first one of these is the effect discussed in Chapter 6, p. 56 in which the orbit of the planet Mercury was observed to precess. This was deduced from the astronomical data of Leverrier in which it was seen that Mercury's orbit about the sun is not precisely periodic, even after the perturbations due to other planets have been taken care of. An aperiodic behavior of a planetary orbit is predicted by Einstein's equations. The precise numerical value for the aperiodic deviation of Mercury's orbit was computed by Einstein from his equations and found to agree very well with the experimental result. This was indeed a triumph for his theory since the observed anomaly was not predicted by any other theory.

The second test of the general theory of relativity was the prediction, in 1916, that as a beam of light comes into the vicinity of a massive body, it deviates from its "straight line" path and appears to bend. In this way, it appears that the theory implies that light has weight! Actually, this effect is due to the fact that light is an electromagnetic signal that travels along the shortest path which connects two electromagnetically interacting bodies. In the absence of any intermediate massive body, such a path would appear to the terrestrial observer as a straight line. However, in the presence of a massive body, Einstein's theory would predict that space–time has curvature and that the shortest distance between two points would not appear to the terrestrial observer as a straight line. It would rather appear to bend in the vicinity of a large mass, such as the sun.

*35. Georg Friedrich Bernhard Riemann (1826–1866)*

This prediction was perhaps more dramatic than the explanation of the precession of Mercury because it was not yet observed in 1916. In the year 1919, the effect was investigated during a complete solar eclipse. Attention was focused on the light that propagates between a particular star and the earth, passing the vicinity of the sun. Einstein's prediction was experimentally confirmed to within a tenth of a second of angle of bending. The result was indeed exciting to many who had hope that the beautiful logic and completeness of the theory of relativity might also be able to stand up under the test of the facts of nature.

A third experimental effect that is predicted by general relativity theory is the following: If electromagnetic radiation should be emitted with a fixed frequency—corresponding, for example, to yellow light—then as this light enters a region of space in the vicinity of a massive body, such as the sun, the frequency of the radiation would change, in this case causing a color change from yellow to, say, orange. This effect is called the *red shift* since the theory implies that light that is in the visible spectrum of frequencies will be shifted toward the red end of the spectrum. The gravitational red shift was actually observed around the beginning of the twentieth century in stellar observations; however, the data was poor at the time and the conclusions were ambiguous. Further experimental investigations with higher resolution were carried out in 1914, when it was established without doubt that there was indeed a gravitational red shift. The exact numerical result was, however, not yet clear. The investigation of this effect was carried out once more in the 1950s with much more experimental accuracy than was ever attainable in the past. The measured value for this effect by this latest study yielded an unambiguous result that is in remarkable agreement with the theoretical prediction of Einstein's theory.

To sum up, the theory of general relativity, when applied to the case of gravitational effects, predicts every result that is successfully predicted by the classical, Newtonian theory of gravitation, as well as successfully predicting three other effects that are not predicted by any other theory. These are (1) the precession of the orbit of a planet, (2) the bending of light rays in the field of a massive body, and (3) the gravitational red shift. Einstein's theory relates to a more general description of gravitation than is described by the classical theory. Thus, the conceptual notions of a relativistic field theory, where the basic elements are continuous entities, triumphs over the conceptual notions of a theory in which discrete things interact at a distance. The two conflicting approaches relate to the old dispute that we have emphasized had existed throughout the history of theories of matter. The triumph of Einstein's theory over that of Newton and Galileo in regard to gravitational forces is then an indication that at least in this domain of natural phenomena, the idea of continuity is more correct than that of a universe which is made up from many discrete things. For the relativity theory implies that the continuous fields that describe mutual relation (oneness), rather than the discrete things of the classical approach, are the elementary entities that most fundamentally describe the universe. It also implies that, in spite of the hints that are given to him by his immediate perceptions, man is forced to a more abstract, underlying theory of the universe than is dictated by "common sense," if a truly logical and complete description is to be attained.

### Attempts at Unification of Gravitation with Electromagnetism

It was Einstein's approach that dictated that all forces in nature—not only those which relate to gravity—are, in fact, manifestations of the most general way to describe space and time. Thus, the theory of general relativity is not only a theory of gravitation; it is rather a theory of interaction. The theory's success in regard to gravitational interaction gave Einstein much encouragement to pursue the approach further in order to see if a more general geometrical scheme, which would go beyond Riemannian geometry, could also include other kinds of forces. His attention was particularly aimed at incorporating electromagnetic forces with gravity. In view of the fact that the observed atoms of matter had been known for several generations (in Einstein's day) to have electric and magnetic properties, he hoped that his program would build a *unified field theory* that would eventually lead to a valid description in the microscopic domain of atoms, as well as to the properties of the astronomical domain—all from the solutions of a common set of fundamental field equations. Einstein felt that his successes with the predictions of gravitational effects were only the first step toward a complete theory of matter. It is unfortunate that he did not complete this program of study in his lifetime.

In retrospect, we see that just as Einstein's physics was revolutionary, so was his entirely objective approach toward science. For he was a strong believer that scientific theories progress most efficiently when total open-mindedness and objective criticism is continually applied to the existing theories. He did not accept any part of scientific principle as absolute truth. He rather believed only in a *provisional truth* in relation to scientific discovery. That is, he believed that scientific principles must always be subject to as much change as will conform with the facts, in a never-ending search for *scientific truth*, and that the direction of this search must be toward a *general theory*. Thus, Einstein was akin to his predecessors in scientific revolution who also, in this way, made giant strides in understanding nature—Faraday, Newton, Galileo, and Aristotle.

We have seen in this chapter that both the special and general forms of the theory of relativity contain successful predictions of the classical theories in addition to other predictions that are beyond the scope of the classical theories. Thus, the theory of relativity has replaced the classical theories of pre–twentieth-century physics as a true generalization. This is not merely a generalization in the mathematical sense; it is, rather, a change that includes the predictions of the older approach but is essentially a change in conceptual content. Thus, the theory of relativity, in relation to the classical theory, is truly revolutionary. But, in line with Einstein's method of science, we should still ask this question: Can one say that the theory of relativity is now firmly established?

Once again, in line with Einstein's approach to science, a positive answer to the preceding question cannot be given by

*36. Ernst Mach (1838–1916)*

any scientific theory. Further, the unanswered problems that still remain in relativity theory make it even more difficult to answer affirmatively. First, the theory of relativity, in its general form, has only obtained confirmation from three independent tests, in addition to its agreement with all of the successful results of Newtonian physics. Further, all three of these tests depend on an approximation to Einstein's field equations. They depend upon the assumption that the gravitational effect of a large massive body, such as the sun, is independent of the time measure of a terrestrial observer, who is at a fixed distance from the sun's center. The gravitational effect must also be assumed to be the same in all possible directions from the sun's center. Further, the way in which this problem was set up is not exactly consistent with the idea of relativistic invariance since it describes a geometry in which a *center in space* (at the source of the assumed spherically symmetric gravitational field) is *fixed*—that is, this center plays the role of a special point in space–time. Since relativity theory requires that, most generally, any point in space–time has no preference over any other, a more crucial test of the theory would come from the predictions of solutions that solve the exact field equations where no special assumptions of symmetry at specific space–time points are made from the outset. These would be predictions that follow from the full form of Einstein's equations, where the matter of the system exists and gives rise to the geometrical field throughout all of space–time. The solutions of this type of equation might relate to the correct structure of the universe, as well as features of nuclear matter. This is because the density of matter is the greatest inside of nuclear matter, as well as in the interior of very hot stars and dense galaxies.

A second, important question that had not yet been answered by relativity theory pertains to the observational fact that gravitational forces are always attractive (according to the known experimental facts), while electromagnetic forces can be either attractive or repulsive. If the theory of relativity does in fact relate to a general theory, then the assumed continuous field properties of space and time should lead to a derivation of inertial mass and, in turn, to the attractiveness of gravitational forces. Thus, it would be expected from this approach that inertial mass is not merely a constant assigned to a bit of matter in order to fit some experimental data. The inertial mass should follow from the fundamental field variables which, in turn, depend on all of the matter in the universe. The notion that the inertial mass of the most insignificant quantity of matter (such as an electron) is in fact a manifestation of all of the matter in the universe (including the most distant stars) was first proposed by Ernest Mach (1838–1916) several years before the onset of relativity theory. Einstein called this assertion *Mach's principle*. It then appears that when one carries the notion of general relativity to

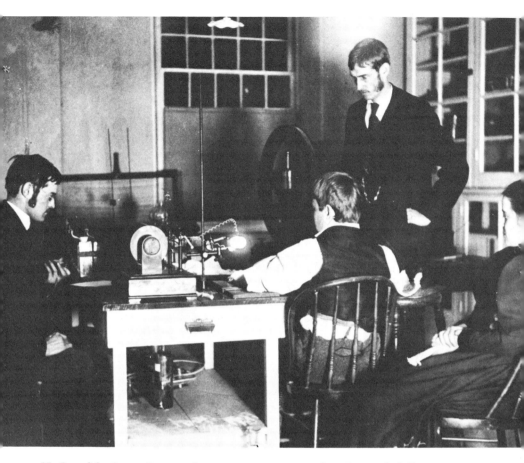

37. *One of the first applications of X rays in America was made at Dartmouth College in 1896.*

its logical extreme, one must incorporate the Mach principle in its definition of inertial mass.

Thus, one might hope to incorporate the Mach principle into the mathematical structure of the field equations in general relativity. The outcome of such a generalization may be a derivation of the inertial masses of the elementary particles of matter. As we have indicated earlier, it was Einstein's hope that if the gravitational and electromagnetic field equations could be unified into one mathematical structure, then the observed, microscopic features of matter might emerge from the solutions of the unified equations. One of these features relates to the mass spectrum of elementary particles. It is interesting that Einstein attributes much of his initial understanding and motivation in relativity theory to his study of Mach. As indicated earlier, Einstein did not complete his program of unifying the gravitational and the electromagnetic phenomena. Consequently he did not accomplish one of his most ardent dreams: to explain the microscopic properties of matter by a single unified field theory.

In the next chapter, we discuss the second revolution of the twentieth century that was evoked to explain the apparent peculiar behavior of matter in the microscopic domain—the quantum theory of measurement. We will discuss its successes and failures and then compare its axiomatic foundation with that of relativity theory to examine the possibility of their mutual coexistence. Finally, some discussion will be given to the outcome of attempts to synthesize the quantum and relativity theories in a relativistic quantum field theory. In the latter approach, an attempt is made to incorporate the quantum and relativity theories by fully exploiting the conceptual notions of the quantum theory, but at the expense of sacrificing some of the basic notions of the complete-continuum field approach of relativity theory. In the last chapter, another approach to this incorporation will be discussed that takes the opposite point of view. That is, a theory will be discussed that fully adopts the continuum field approach and fully exploits the conceptual idea that is built into the general relativistic approach, but at the expense of sacrificing some of the basic notions to the quantum theory. The conflict between the latter two approaches is the same one that has been emphasized throughout the book and has persisted since earliest times in the history of science—the conflict between atomism and continuity as fundamental aspects of matter.

# 8

# Quantum theory of measurement

The apparently peculiar behavior of microscopic systems (Chapter 6) leads to the notion that the energy, momentum, and the angular momentum of the smallest bits of matter do not have a continuous distribution of values. These properties appear from the measurements (in blackbody radiation, the photoelectric effect, atomic spectra, and so on) to have only particular numbers assigned to them. Thus, Planck and Einstein concluded near the beginning of the twentieth century, that the conserved quantities in a physical system must be "quantized" in the microscopic world. In accordance with this approach, Bohr explained the spectral lines from radiating atoms by asserting that the electrons, which orbit about the atomic nucleus under the influence of the Coulomb electric force, do so only with respect to particular (discrete) values of energy and angular momentum. Thus, according to this model, given any two electron orbits in an atom, there can only be an integral number of other orbits, or none, in between them. This is in contrast to the properties of the planetary orbits in the solar system. For in the latter case, as the surface of a planet wears away in its motion through atmosphere that moves at a different speed, the planet does indeed gradually change its energy and angular momentum in a *continuous fashion*. The electron orbit, according to

Bohr's model of the atom, can only change by making a discrete "jump," accompanied by the creation or annihilation (depending on whether the final orbit corresponds to less or more energy than the original one) of a quantum of electromagnetic energy that is called a "photon." Such a picture of discrete orbits also resolved the difficulty that would arise if the motion of these charged particles were treated in a classical way.

As we discussed earlier, an accelerating charged particle of matter must emit energy in the form of radiation. This would imply that a rotating electron should *continuously* lose energy as it would spiral right into the nucleus. Thus, the model of the atom, that was originally deduced from the analysis of Rutherford's scattering experiments, could not be stable if the usual description of electrodynamics were applied. However, the electrons would be forced to stay in their orbits without losing energy if the energy values that were available to them were discrete, rather than continuous. Thus, the *postulation* of quantized energy and angular momentum provides a model of the atom that forces it to be stable! It also gave rise to numerical predictions for the energy levels of atoms that were in agreement with the spectral data that was then available. Nevertheless, such a postulate did not seem to be rooted in anything else and it seemed somewhat *ad hoc*. Thus it was not quite satisfactory to the scientific community, even though it did give some agreement with the actual data!

Recall further that the experimentation on electron diffraction from crystals led Davisson and Germer, in 1928, to the conclusion that a massive particle of matter can behave like a wave under the proper experimental conditions of measurement. In other words, the electron can be seen to have a wavelength and a frequency value that is associated with its "particle" features, such as momentum and energy. Thus a *group* of electrons under these same experimental circumstances would interfere with each other in a constructive and destructive fashion just as ordinary waves would do. These results then gave a great deal of credence to de Broglie's earlier (1924) hypothesis that related a particle's momentum to its wavelength according to the simple equation that uses Planck's constant.

Knowing that there is a matter field (or "wave function") associated with the bit of charged matter that is called electron, the next step was to find the law of nature that precisely determined this field. The hint as to the structure of the equation that expresses this law comes from the observation that the wave function for the electron will add to another wave function for the same electron, *in a linear fashion*, to give a third wave function that also describes the same electron. This is a feature of linear-type waves—such as electromagnetic waves or the ripples produced by a disturbance on the surface of a pond.

This is meant in the following way: Suppose that $W_1$ represents the height of a wave that is moving through space. It would generally depend on the spatial location, $r$, relative to some point in space (say the location of a particular observer) and it would depend on the time, $t$, relative to some initial time. The way in which we express the fact that the height of the wave depends on $r$ and $t$ is to write $W_1(r,t)$. $W_1$ is then said to be a "function of" $r$ and $t$. Now, $W_1$ is assumed to be the solution of some equation (the law of nature which we seek). It is called a "wave equation" since $W_1$ describes a wave. If $W_2$ is a different solution of the same equation, then if this equation is linear, the sum $(W_1 + W_2)$ is a third solution that solves the same wave equation. A linear equation is one that depends on solutions to no higher than the first power.

The assumption that the sum of any number of solutions of the electron wave equation is also a solution of the same equation is based on experimental observations. The first indication of this property of electrons came from the electron diffraction experiments of Davisson and Germer and of G. P. Thomson. The interference of the electron waves then implied that the underlying equations that describe matter in the atomic domain are linear equations. The experimental evidence on the additivity of the electron waves to form a new wave that also solves the underlying equation was then asserted as a fundamental principle—the *principle of linear superposition.* Starting from this point, then, physicists had a strong hint about the mathematical structure of the law of nature that must describe the atomic domain of matter.

The first successful form of the equation for electron waves was discovered by Erwin Schrödinger (1887–1961). The so-called *Schrödinger wave equation* was constructed in a way that includes the mechanical properties of the observed matter in terms of its wave nature. The resulting equation describes the way in which the electron wave function changes in time as a consequence of its particular mechanical features. Just as Maxwell's equations provided a quantitative and precise expression of the field concept of Faraday to describe electromagnetism, so Schrödinger's equation did this for de Broglie's wave–particle interpretation of matter.

Around the same time that Schrödinger's equation was developed (in the mid-1920s), two other theoretical physicists were viewing the wave nature of matter in a somewhat different way. These men were P. A. M. Dirac (1902–   ) and Werner Heisenberg (1901–   ). This study started with Heisenberg's recognition that the numbers that correspond to the observed distribution of spectral lines of radiating atoms could be arranged in a two-dimensional "matrix" with a special kind of order. The complete set of such matrices, which would describe all possible observables of the atoms (energy, momentum, and so on), obey a certain kind of algebra. That is, these matrices, as units by themselves, add, subtract, and multiply according to a prescribed set of rules. These rules, applied to the set of matrices that were considered, are called *matrix algebra.* Such an algebra was well known to mathematicians since the middle of the nineteenth century. It was Heisenberg's discovery that some of the essential features of matter in the microscopic domain can be accurately described in terms of a matrix algebra.

One of the very interesting features of matrix algebra that is not obeyed by the algebra of ordinary numbers is that the product of any two matrices generally depends on the order in which they are multiplied. That is, if $a$ and $b$ are any two ordinary numbers, it is *always* true that $ab$ and $ba$ are the same number, or

$$ab - ba = 0.$$

On the other hand, if $A$ and $B$ are any two matrices (let's say that they are square arrays with the same number, $n$, of rows and columns each), then according to rules of multiplication of matrices, when they are multiplied, another matrix, $C$, results from the product $AB$. Yet a different matrix, $D$, would result from the product $BA$. In special cases, $C$ could be equal to $D$, but it is generally not so, that is,

$$AB - BA \neq 0,$$

where 0 represents an $n$ x $n$ matrix of zeros. For the purposes of this discussion, it is not important to specify the rule for matrix multiplication. The important point is the fact that the product of any two matrices (of equal dimension) depends on the order in which they are multiplied.

Dirac observed that the dependence of the product of two matrices on their order of multiplication is also a feature of the product of two linear operators. The term linear operator here refers to the mathematical way of representing an instruction. Such an instruction is given to a mathematical expression that describes an existing state of matter and it usually orders some specific sort of change to take place. Let us illustrate this with a simplified example. Suppose that the underlying state of matter is a bare foot. Let us represent this foot, mathematically, by the symbol f. Now let the instruction to the foot be: *put on a stocking*.

We might represent this instruction with the symbol $\widehat{S}$. (We use the symbol $\widehat{\phantom{x}}$ to remind the reader that $\widehat{S}$ is not a number, like 5 or $\pi$, but that it is, rather, an "operator," or an instruction. Such an entity would be meaningless without something to operate on (to give an instruction to) such as the function $f$ in this example.)

The operation of putting a stocking onto a foot is then represented symbolically as $\widehat{S}f$. The net effect of this operation is the appearance of a different sort of foot—one with a stocking on it. If we represent the latter by the symbol $g$, then the effect of putting a stocking onto a foot can be symbolized by the equation:

$$\widehat{S}f = g.$$

In other words, this equation states: The act of putting a stocking onto a foot results in a new type of foot—that which is covered by a stocking.

Suppose now that $\widehat{B}$ is the operator which corresponds to the instruction: *put on a boot*. If $h$ is the symbol for a foot with a stocking on it that has been covered with a boot, then this operation can be represented symbolically with the equation

$$\widehat{B}g = h.$$

Since $g = \widehat{S}f$, the preceding equation can also be expressed in the form

$$\widehat{B}\widehat{S}f = h.$$

Now consider that the order of operations has been inverted. First we must put a boot onto a bare foot, represented symbolically as $\widehat{B}f$. Then we apply $\widehat{S}$ to the result (that is, place a stocking over the foot that is already covered by a boot), yielding.

$$\widehat{S}\widehat{B}f = k.$$

Clearly $k$ is not the same as $h$. The symbol $k$ represents a bare foot covered first with a boot and on top of this, a stocking; the symbol $h$ represents a foot covered first with a stocking and then with a boot (the way we normally do it!). We can express this as follows:

$$(\widehat{B}\widehat{S} - \widehat{S}\widehat{B})f = (h - k).$$

Since $(h - k)$ is not equal to zero, it follows that $(\widehat{B}\widehat{S} - \widehat{S}\widehat{B})$ is not equal to zero. That is $\widehat{B}\widehat{S}$ and $\widehat{S}\widehat{B}$ are different sorts of instructions. In this case, $\widehat{B}$ and $\widehat{S}$ are said to be "noncommuting operators." They are called linear operators because the instructions themselves ($\widehat{B}$ or $\widehat{S}$) are independent of the specific properties of the foot—such instructions could have been given to a large foot or a small foot, a man's foot or a woman's foot, etc., without altering the form of the instruction itself. If, for example, $\widehat{S}$ had been dependent on the foot itself (that is, if $\widehat{S}$ would depend on $f$) then $\widehat{S}f$ would depend on $f$ to a power other than unity (it may depend on $f^2$ or $f^3$ or $f^{1/2}$, etc.). In the latter cases, $\widehat{S}$ would be called a nonlinear operator. It is important that Heisenberg's matrix representation of atomic phenomena relates specifically to linear operators only.

Thus we have seen that the application of more than one linear operator has a feature that is in common with the multiplication of matrices. This is the fact that the final product depends on the order of application (multiplication) of these entities. The idea follows, then, that there might be some sort of correspondence between linear operators and matrices. If this is so, then how must we interpret these linear operators, which will correspond to the matrices that were used in Heisenberg's description of the atomic spectral lines? Dirac answered this question by noting that if the field variable $f$ is to represent, in some way, the sum total of attainable knowledge about the physical system that is under investigation, then the linear operator $\widehat{L}$ might then relate to an *act of observing* the system. Thus, if we should be able to list the complete set of properties of a physical system that can possibly be measured, then there would correspond a *complete set* of linear operators $\hat{L}_1$, $\hat{L}_2$, . . . and a complete set of values $v_1$, $v_2$, . . . that are the possible outcomes of the measurements represented by the preceding set of linear operators.

With this interpretation of the linear operators as representing measurements, what is the meaning of the fact that it sometimes makes a difference if one of the properties of the system is measured first, and then the second property is measured, or if they are measured in the reverse order? After all, we have never encountered such a situation before in any of the measurements of the properties of matter. If, for example, I should measure the position of the planet Mercury and then make a measurement that would imply its momentum at that position, or if I should first measure the momentum of the planet and then make a measurement leading to its position at the place where the momentum was determined, it should make no difference. That is, according to Newtonian physics, there is nothing that prevents me from determining the precise momentum and location of the planet at all times and with as much precision as I like. Another way of stating this is to assert that in classical physics, all of the physical properties of a system are "predetermined"; *they are independent of whether a measurement is made or not.* According to the interpretation of linear operators as they relate to the measurement of the properties of atomic systems, the physical properties of the system are not predetermined. In other words it does indeed make a difference in some cases as to the order of measuring two of these properties. Thus, according to this view, not all of the properties of a system can be determined at the same time with arbitrary accuracy. The reason given is that, in a certain way, these two properties "interfere" with each other.

If one should try to measure the momentum of an electron (instead of the planet Mercury) the apparatus would be said to disturb the particle in a way such that its precise position could not be deduced where the momentum had been determined. The more accurately that we try to determine the momentum of the electron from a measurement, the less accurately can we know its position at the same time. This assertion is the "Heisenberg uncertainty principle." If the position measurement is symbolized by the linear operator $\hat{x}$, the momentum measurement is symbolized by the linear operator $\hat{p}$, and if $f$ represents the wave function which would imply all of the possible measurable properties of the electron, then the preceding conclusion may be symbolized as follows:

$$(\hat{x}\hat{p} - \hat{p}\hat{x})f \neq 0.$$

If this expression is indeed non-zero, then what is it? In the full development of this theory (quantum mechanics) it is found to be equal to a constant number multiplied by the wave function itself, that is,

$$(\hat{x}\hat{p} - \hat{p}\hat{x})f = (\text{constant})f.$$

When this description of the measurement of microscopic systems is put into practice, the comparison with the outcome of actual experimentation leads to the "constant" above. It turns out to depend on the constant which is precisely equal to $h/2\pi$, where $h$ is the same Planck's constant that was determined earlier in a host of atomic experiments and related then to the "old quantum theory."

As in the formulation of the Schrödinger equation, these "measurement operators" (rather than any "differential operator") are *linear* since they do not depend on the function $f$ that is being acted upon. The reason is that $f$ is supposed to relate to a field that contains the predictions for the properties of matter, while the linear operators represent the action of the measuring apparatus in viewing this matter. That is, the description of the measuring apparatus is assumed to be independent of that of the matter that is observed.

38. *An experiment conducted by Aldine in 1803 to determine the speed of electric flow along a wire stretched across the bay of Calais.*

Suppose now that one of the possible linear operators (out of the complete set that is available to us) corresponds to the measurement of energy. This particular linear operator, will be denoted by $\widehat{H}$. If $\widehat{L}$ represents the measurement of any other property of the system, then it turns out that

$$(\widehat{H}\widehat{L} - \widehat{L}\widehat{H})f(x) = (\text{constant})\left(\frac{d\widehat{L}}{dt}\right)f(x),$$

where the constant again depends on Planck's constant $h$ and $d\widehat{L}/dt$ is a symbol that represents the time rate of change of the operator $\widehat{L}$. With this formalism, Dirac and Heisenberg attempt to describe the data with an equation that tells us how some linear operator (that corresponds to the measurement of some physical property of the system) changes with time. The latter time rate of change *operates on* the field variable $f$ that is defined at a particular space–time point. As we have indicated earlier, the latter variable, called the *state function,* or the wave function, implies a particular set of measurable properties of the observed system.

The Dirac–Heisenberg description should be contrasted with the Schrödinger equation in which the linear operator is fixed while the time rate of change of the wave function $f$ is determined. The latter description of a microscopic system is called the "Schrödinger representation," while the former is the "Heisenberg representation" for the *quantum mechanical equations*. The implication of calling each of these a different "representation" is that they are each different ways of describing the same thing. This is indeed the case, as was shown in the early stages of these investigations by Schrödinger. For, after the solutions for both of these types of equations are found, they can be used in a prescribed way, that is consistent with the forms of the respective equations, to predict the numbers that are supposed to be the actual outcome of the measurements which are performed. When this is done, it is found that the actual mathematical expressions which are to be used to compute the numbers (to be compared with the experimental data) are identical in every way for both representations of quantum mechanics. The reason has to do with the one-to-one correspondence between the matrices in Heisenberg's approach and the linear operators that are used to construct the wave equation in Schrödinger's approach. The resulting formalism, *quantum mechanics*, has given predictions for the properties of microscopic systems that are in extremely good agreement with the numerical results of the experiments. With this approach, results were obtained that can be understood from a logical point of view, and yet are entirely incompatible with the classical theory—or even with the "old quantum theory" of Planck and Einstein. There are other features of the quantum mechanical view of measurement whose logical consistency is still being debated at this time. Some of this argumentation will be given later on in discussions of the paradoxes of Einstein, Rosen, Podolsky, and of Schrödinger.

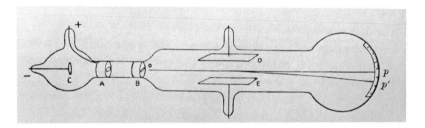

*39. A stream of cathode rays in an evacuated tube.*

It should be pointed out that while the Schrödinger wave equation and the Heisenberg equations of motion *both* triumphed in predicting the facts (because they were mathematically equivalent), the original interpretations of their respective formalisms by Schrödinger and de Broglie on the one hand, and by Dirac and Heisenberg on the other, were quite different. The difference was, in fact, rooted in the continuous versus discrete aspect of the fundamental description of matter. Schrödinger wanted to view his equation as a "field equation," whose solutions describe "matter waves." The form of his equation then described the way in which these waves must depend on spatial coordinates and how they must propagate in time. This is analogous to the way in which one would want to describe the ripple of the surface of a pond from the equations of hydrodynamics. With the Dirac view, on the other hand, this "wave function" is not at all to be taken as a real wave. It rather *represents* the "state" of a bit of matter in the sense of "state of knowledge" on the part of the observer. Specification of this function would permit, according to a prescribed set of rules, a calculation of all of the measurable properties of this bit of matter when it is in the particular state. Thus, Dirac would prefer to call this a "state function" rather than a "wave function." In both cases, however, no claim is made that this function is directly observed! It is only a mathematical function that must be *used* to calculate the numbers which are, in turn, to be compared with the observations.

Whether one is using the Schrödinger or the Heisenberg representation, the actual mathematical expressions that are used to calculate the observables from the "state functions" have the same form that would be used conventionally to compute an *average value* of some property. Thus, one might say, according to the mathematical language that is used, that the formalism of quantum mechanics is used to deduce the average values of the properties of matter. These are, in turn, the numbers that are to be compared with the outcome of actual measurements which are supposed to entail looking at matter with the use of laboratory instruments that are made up of a very large number of atomic systems. But the theory then goes on to say that, in contrast with the approach of statistical mechanics (which also attempts to derive average values of a very large number of atomic systems that constitute an ordinary quantity of matter), it is not meaningful to impose a precise, predetermined set of values, which is later to be averaged by the formalism.

Let us consider an example that contrasts these two conceptual approaches to statistics. An example of an average value in classical mechanics is the center of mass of a distribution of matter. One arrives at this quantity by multiplying the mass of each of the parts of a given system by its distance from a fixed point, and by dividing the sum of all such quantities for the whole system by the sum of all masses in the system, that is,

$$R_{c.m.} = \frac{(m_1 R_1 + m_2 R_2 + \cdots + m_n R_n)}{(m_1 + m_2 + \cdots + m_n)}$$

where $n$ is the number of parts that constitute the system and the quantity $R_j$ stands for the distance from the particle of mass, $m_j$, to some fixed point. In this case, it is meaningful to define the values $m_1$, $m_2$, ... $m_n$ and $R_1$, $R_2$, ... $R_n$ with arbitrary precision at the same time—even though one may not actually measure these values separately. These are a "predetermined" set of values—they are independent of whether or not a measure-

*40. Erwin Schrodinger (1887–1961)*

ment will be made. On the other hand, the separate parts that are added together in the quantum theory of measurement, to compute an average value, are *not defined* in an operational way as are the masses of the separate parts in the above example. In this example, the masses of each of the separate parts "weights" their distances to a fixed point. That is, a more massive body at some distance makes more of a contribution to $R_{c.m.}$ than does a less massive component at the same distance. On the other hand, it is the (*square* of the magnitude of the) wave function $f$ that would "weight" a position element to compute an average distance in the quantum theory. But the wave function has nothing to do with an independent constituent bit of matter according to the quantum theory of measurement. Rather, $f$ is supposed to relate to the interference between a measuring apparatus and some matter that is observed. In contrast with the classical theory, there would be no description at all if there was not a measuring apparatus in the system. There does not exist a "predetermined" set of values for the properties of matter, according to the quantum theory. Thus, the "averages" that one talks about in this approach are as much as one can ever say about the physical system—one can never decompose the average value into precise parts, as one does in the classical theory.

In summary, the quantum theory of measurement, which was expounded by Dirac and Heisenberg, also studied earlier and given philosophical justification by N. Bohr, takes the view that the properties of any physical system are not predetermined with all values simultaneously assigned precise values. The measuring apparatus must be involved in the *definition* of the physical properties of a system. It is allowed to probe the value of any particular property *as closely as it pleases*—but only at the expense of limiting the accuracy to which the values of other properties can be known from this measurement. The properties of matter, according to this point of view, depend on the way in which they are measured, and thus matter is described in a *subjective* rather than an *objective* way. Since the properties of matter are not predetermined according to the quantum theory of measurement, it is a *nondeterministic* theory.

It should be emphasized that the nondeterministic aspect of this theory is not in the same sense as it is meant in the classical, statistical problems. In the latter case, one only says that because of our personal inability to devise a measuring apparatus that is accurate enough to measure all of the values of the properties of a system, the statistical method will be utilized to predict their average values. This approach is still *deterministic* at the fundamental level since a predetermined set of values for these properties (to arbitrary accuracy) is indeed assumed to exist. The quantum theory of measurement is nondeterministic because it denies the existence of a predetermined set of values for the properties of matter.

The introduction of such a nondeterministic theory of matter in which the measuring apparatus must play an essential role in the fundamental description was indeed a revolution in the history of science. In contrast to the revolution of relativity theory, however, this one did not seem to be based on any of the preceding theoretical developments. Its strength was primarily in the very large amount of agreement that was obtained between the predictions of the equations of this theory and the experimental facts relating to the atomic domain.

## Paradoxes in the Interpretation of the Quantum Theory of Measurement

The equations of the nonrelativistic theory of quantum mechanics were consistent from the mathematical point of view. The equations also predicted a great deal that was in agreement with the experimental facts and also resolved earlier problems that resided in the "old quantum theory" of Planck and Einstein. However, the nondeterministic approach and the subjective interpretation of the fundamental properties of nature, according to this theory, were not at all acceptable to many of the leaders in the scientific community. Both Einstein and Planck—the founders of the old quantum theory—were quite opposed to the ideas of the new quantum theory. It is also interesting that both Schrödinger and de Broglie, two of the originators of the mathematical form of the new quantum mechanics, were quite opposed to the interpretation of their equations that was given by Bohr, Heisenberg, and Dirac. The opposition of these scientists was based on both the aesthetic desire for a complete description of nature, that is independent of the particular type of measuring apparatus that happens to be used to observe one effect or another, and on their belief that the new quantum theory of measurement was lacking in logical consistency.

Thus, in their writings, both Einstein and Schrödinger could not accept a theory which tells the inquirer that he is not able to ask any meaningful questions about an underlying predeter-

be measured, the less accurately could the values of other properties of $A$, say $x_A$, be known. On the other hand, since the states of motion of atom $A$ and atom $B$ are still correlated (according to the original correlation that they had when they were bound as a molecule $AB$), it also follows that an accurate knowledge of $P_A$ would imply an equally accurate knowledge of its correlated variable $P_B$ of the atom $B$. Similarly, by choosing to initially measure the variable $x_A$ to arbitrary accuracy, $x_B$ could then be known to the same accuracy. In other words if you measure one property of atom $A$ to a certain accuracy, you will know the same property of atom $B$ with the same accuracy. We see then that the observable properties of the atom $B$ can be determined to arbitrary accuracy without disturbing $B$ at all! Thus, the values for all of the properties of the atom $B$ are predetermined. On the other hand, the quantum theory asserts that the state function for $B$ relates only to the response of a measuring apparatus to the motion of the atom $B$. With this view, all of its values cannot be predetermined before the measurement is carried out.

Thus, EPR adherents were able to draw two opposite conclusions from the same initial assumption about the interpretation of the state function (the solution of the quantum mechanical equations). They concluded that the paradox will only disappear if we are willing to accept the state function in terms of a weighting function that is used only to determine the average properties of a large number of atoms or elementary particles. In their view, then, the quantum, mechanical solutions relate to an incomplete description of a system that contains a large number of things. The state functions are then to be used in the same way that the distribution function is used in statistical mechanics, to find the average properties of a gas of atoms. Thus, the success of quantum mechanics in deriving the correct average values for the properties of matter in the microscopic domain was taken by Einstein to mean that the latter mathematical equations are not more than a way of averaging over an ensemble of atoms, which indeed have an underlying, complete, and predetermined description.

The EPR paradox caused a great deal of concern in the scientific community since it cast serious doubt on the logical consistency of the then-accepted interpretation of the quantum theory of measurement. The chief rejoinder in defense of the quantum approach came from Bohr. Before discussing his answer, however, we will briefly outline the cat paradox of Schrödinger. Bohr's answer attempts to rebut both of these paradoxes at once.

Schrödinger discussed the following hypothetical situation. Consider a large box that has two compartments that are separated by a partition with a small hole in it. Into one compartment we will place a piece of radioactive metal, radium, for instance. In the other compartment we will place a live cat and some equipment. The equipment consists of a Geiger counter (which is placed in front of the hole in the compartment wall), a jar of poison gas, and an electrically operated hammer that is aimed at the jar. The equipment is arranged so that if an alpha particle given off by the radium in the one partition should travel through the hole, it would go into the Geiger counter. An electric current would then be caused to flow through a circuit that includes a magnetic solenoid. The electric current through the latter component would cause a mechanical lever to move the hammer which would then crack the jar and release the poison gas. In summary, if the alpha particle should go through the hole in the wall, the cat in the other compartment would die! If no alpha particle should go through the hole in the wall, the cat would live! We see then that there is a one-to-one correspondence between the states of motion of the alpha particles that are emitted by the radioactive radium and the states of life and death of the cat.

According to the quantum mechanical interpretation, so long as a measurement is not made, the alpha particle can be in any state of motion at all. The theory then says that one must describe it in all possible states at the same time, with equal probability. It is only after the measurement has been made that one of these states of motion is "projected out" and deduced from the readings of the apparatus. But this means, in Schrödinger's example, that so long as one does not look into the box, the cat must be in a superposition of the states of life and death, that is, the cat is both dead and alive at the same time! By "looking" we do not necessarily mean that a human eye is looking. We mean the measuring device (of any type) is "looking" at the cat. It is only after one looks into the box, according to this approach, that the cat will assume one of these states with certainty—after one looks into the box he will see that the cat is either dead or alive! Schrödinger then concluded that since the life or death of the cat should not actually depend on whether or not someone looks at it, and since the cat cannot be both dead and alive at the same time, before it is perceived, the *subjective* description of the real situation that is proposed by the quantum theory of measurement is not logically consistent.

One further feature of the quantum mechanical view that is highlighted in the cat paradox is the following: It is contended by the quantum approach that while the mechanics of microscopic matter (like the alpha particle) must obey the rules which assume that there does not exist a predetermined set of values for its properties, that its actual values of position, momentum, and so on, have something to do with the way in which these properties are observed; large macroscopic objects (like a cat) must obey the rules of classical physics. In the latter description, the system is precisely described with a predetermined set of values for all of its properties which are independent of any sort of measurement. According to our earlier discussion regarding the Dirac–Heisenberg approach, the nondeterministic character of microscopic matter is a consequence of the process of measuring any of its properties and the interference that is thereby set up with the values of some of its other properties. The inter-

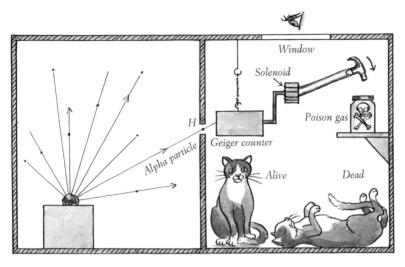

*41. Schrodinger's Cat Paradox*

ference comes about when the energy and momentum of a photon (or other particle) that "looks" at the alpha particle in the
measurement is of the same order of magnitude as the energy
and momentum of the alpha particle. The large classical object
(like a cat), on the other hand, is too massive to be "disturbed"
in this way by its collision with some photons. However, we have
seen in Schrödinger's example that the states of the large object
(the cat) must indeed be described in the quantum mechanical
way under the special circumstance that related to its coupling
to the microscopic object (the alpha particle). Thus Schrödinger
challenged the idea that there was a sharp separation between
the classical description of the large scale quantities of matter
and the quantum description of microscopic matter.

While Niels Bohr addressed his rebuttal to the argument of
Einstein, Podolsky, and Rosen, it also applied to Schrödinger's
argument. Bohr admitted that so long as the state function is
taken to relate to only a part of a coupled system, the paradoxes

would remain. However, he did not interpret the state function as relating to a partial system. Rather, it has to do with the response of a measuring apparatus to a physical system as it exists when it is being studied. In the EPR example, the state function relates to the response of a classical measuring apparatus to the whole molecule $AB$; or, an entirely different state function would relate to the response of an apparatus to a single atom ($A$ or $B$). Thus, Bohr contended that it is not meaningful to trace the history of the atom from the time when it was bound in a molecule to the time when it became unbound. The time coordinate, in this description, refers only to the time rate of change of the state function as a consequence of a given measurement. Similarly, the state function in Schrödinger's example relates to the case of life, or the case of death. It does not have to do with the history of the physical system from life until death. The state function has only to do with an existing state of affairs. Thus, Bohr would contend that every time a different measurement was made, one must start anew without any imposed knowledge of an exact history of the system. All that one can say about the observed matter, according to this view, is that the measurements imply that certain *possible* states of motion must be intrinsic and that certain states occur with particular probabilities as compared with other states of motion.

In this way, the paradoxes of Einstein, Podolsky, and Rosen and of Schrödinger were removed. But it was only done at the expense of assuming that matter does not have a *complete description*, in the sense of the determinism of classical physics or the field physics of Faraday and Maxwell. That is, the nondeterminism that we have described as inherent in the quantum theory of measurement, must provide, according to Bohr, as complete a description of matter as one can possibly get. As we have indicated earlier, Einstein and Schrödinger disagreed with Bohr's conclusion. They felt that a completely determined description of nature does indeed exist, even though it has not yet been given explicitly, and in spite of the mathematical successes of non-relativistic quantum mechanics. They anticipated that such a complete description must necessarily go beyond the quantum mechanical way of predicting the possible outcome of measurements in a way that would entail predetermined values for all of the properties of any quantity of matter—and independent of the way in which these properties should be measured. Thus they viewed quantum mechanics as an intermediate stage in the development of physics that has been useful to describe the average values of the properties of atomic matter, but awaits a more complete description of the underlying nature of matter.

Should one agree with Bohr's *philosophical assertion* that a precise knowledge of a microscopic portion of matter is necessarily limited *in a fundamental way*, then the mathematical variables that appear in the equations of quantum mechanics must incorporate this uncertainty in the variable quantities that are described. With this interpretation, there is no paradox. The logical difficulty occurs only if one should insist (with Einstein and Schrödinger) that the state functions refer to individual systems, rather than to the response of a measuring device to an existing physical system.

While Einstein and Schrödinger were unwilling to accept Bohr's assertion, they could not reject it on logical grounds since an assertion is neither true nor false. It is, rather, an *axiom* that is to be tested according to certain rules which provide the means of drawing conclusions. These conclusions are eventually to be compared with the experimental facts. Nevertheless, the question of logic enters the discussion again as soon as the attempt is made to unify the ideas of the quantum theory with those of relativity theory. Such a unification, which is necessary if one wishes to describe microscopic bits of matter that have very high speeds (close to the speed of light), will be discussed in the following chapters.

# 9.

# A comparison of the quantum and relativity theories

Thus far we have discussed the two major scientific revolutions that have occurred in this century in the search for a consistent theory of matter. We have seen that the theory of relativity provides a theoretical framework that not only predicts the same results that were successfully predicted by the classical theories, but that also explains a host of other experimental results that do not have any satisfactory explanation in the classical theory. Similarly, the quantum theory of measurement has successfully predicted a great many of the features of microscopic matter that were not predicted previously by the classical theories. The theory of relativity approaches the classical form if we assume that the speed of light is replaced by infinity in comparison with the relative speeds between different frames of reference, say between the "observer" and the "observed." Similarly, the quantum mechanics that describes nonrelativistic, microscopic matter takes the form of classical mechanics if we consider Planck's constant to be zero, compared to the energy of a measured system multiplied by the time of measurement. This product of energy and time is called action in mechanics. The assumption that Planck's constant is zero is a very good approximation whenever we deal with the behavior of large scale quantities of matter. Thus we see that both the theory of relativity and the quantum theory adhere to a correspondence principle.

Relativity theory and nonrelativistic quantum theory cover a much wider range of observations than do the earlier classical approaches to theories of matter. Thus, they replace the earlier theories as generalizations. It is important to understand that this is more than a mathematical generalization. On the contrary, the important changes that have taken place are in the conceptual content of the respective theories. It is the conceptual change that leads to a new mathematical formalism which, in turn, approaches the forms of the classical theories in the particular limits discussed above.

It is the change in conceptual approach, rather than a change in the mathematical language, that constitutes a scientific revolution. Granted that a revolution in thought has occurred in these two phases of physical description, the reader should take note that the older theories were never completely abandoned. The correspondence principle has provided a framework of language to which the general theory must necessarily conform. That is, the theory must fit within the limits that were successful in classical theory. In this way, the knowledge gained in the search for a general theory of matter has always been added to previously acquired knowledge.

It is interesting to note that for the first time in the history of science, the twentieth century has witnessed the simultaneous birth of two scientific revolutions. At their inception, the experiments that were explained by the theory of relativity and those explained by the quantum theory were mutually exclusive. So long as one did not have to do with the other, the two theories could be considered to be explanations for separate types of experimental evidence—at least by those who would be satisfied with a different formula for each observation. On the other hand, from the point of view of theoretical physics, this state of affairs could not be acceptable in the long run unless it could be shown that at least these two theories were not in disagreement with each other from the point of view of their concepts. This follows from the belief in a general theory that could cover all physical phenomena.

Besides this reason, there was also another compelling motivation for attempting to join the quantum and relativity theories. This had to do with the experiments in the late 1920s and afterward, dealing with atomic and nuclear physics. For one thing, it was observed that the spectral lines of radiating, excited atoms actually involve the coupling of an atom with an entity. The entity is described by the electromagnetic field which must necessarily be described within the theory of relativity. The atom, on the other hand, is described by the quantum theory. Secondly, the development of nuclear accelerating machines (atom smashers) led to experiments which involve the coupling of microscopic matter with microscopic matter whose relative speeds are close to the speed of light. To describe these observations necessarily led to the requirement of fusing the theory of relativity with the quantum theory.

The fascinating and provoking aspect of the appearance of these two scientific revolutions at the same time in the history of science is that while separately they are logically consistent, together as one general theory they would introduce incompatible ideas as a basis for this theory. Because of this unhappy state of affairs, it is not possible to join the two theories unless the conceptual parts of one that are not compatible with those of the other are removed. Let us then outline, in somewhat more detail, specifically what are the compatible and the incompatible features of the quantum and relativity theories.

### Discreteness versus Continuity

The most striking point of departure in these two theories is one that is very old in the history of science. This is the difference between discreteness and continuity as basic features for the underlying, fundamental description of matter. The question as to the choice between these two descriptions of matter has persisted throughout the entire recorded history of physics. We have seen, in our story of a search for a theory of matter, that in the various periods of history, the advocates of one of these views would seem triumphant for a while, only to be demolished by further progress in experimental science that could only be explained with the opposite view. The twentieth century is unique in its development of two simultaneous revolutions, one based on the hypothesis of discreteness (quantum theory), and the other on continuity (relativity theory), to describe matter in the most fundamental way.

How do we know that each of these theories is in fact opposing the other in this way? To answer this question, let us first consider the quantum theory of measurement. It was assumed in the old quantum theory of Planck and Einstein, as well as in the newer innovation of quantum mechanics, that all measurable properties of the smallest bits of matter must necessarily have a discrete set of values. Just as the integers, 0,1,2,3, . . . are a discrete set of things, so it is assumed in the quantum theory that the properties of matter, such as the energy content, can occur only in a discrete sequence of values, $E_0$, $E_1$, $E_2$, $E_3$. . . . Just as there are no integers between 2 and 3, so the quantum theory contends that there can be no *measurable* energy for a microscopic system between any possible values of energy, say $E_2$ and $E_3$.

To illustrate the concept of discreteness, consider the second hand of a watch as it rotates in discrete jumps, corresponding to the individual seconds. The functioning of the second hand in this type of a watch is due to the rapid impulses of momentum that are imposed at every second. If momentum is indeed quantized in this way, then the workings of the watch would not permit one to arrange the motion of the second hand to jump every one-half second, or every one-hundreth of a second, or to move continuously (where the motion is *smooth*). The workings of the watch would only allow a jump to occur every second, since, according to the hypothesis, this is the only possible momentum that could push the second hand of the watch through a distance on the face of the watch that exactly corresponds to one second.

The feature of discreteness appears in another important way in the quantum theory. For it is assumed from the outset, in this theory, that the world is made out of a large number of discrete things (the atoms of matter). But, in contrast with the earlier approaches to atomism, the quantum theory does not agree that there exists an exact description of the whereabouts of these discrete quantities of matter. Still, this theory does use, as the basic elements of its language, the variables that have to do with the states of motion of these individual things. In other words, even though we may not know where atoms are, we can still determine their motion, on the average.

As we discussed earlier, the quantum theory predicts the measured values of some property of microscopic matter by disturbing the motion of the "thing" at the time of measurement. This causes it to change its path and go into its other available states of motion. The description of the motion of the "thing" is indeed more complicated here since it is contended that one cannot know with certainty which particular state of motion, out of a whole spectrum of available states, the bit of matter possesses at any particular time. We have seen that one can only talk about the average values of properties of matter and the probabilities that particular states of motion will be transformed into other states of motion. Even though this description is more sophisticated and mathematically complex than the Newtonian theory of moving matter, it is still the individual, discrete "thing" in the quantum theory, as it is in the Newtonian theory, that forms the basic element of matter.

In contrast with the approach of the quantum theory, the

conceptual foundation of the theory of relativity is based on the notion of *continuity*. This can be seen in Einstein's original meaning for the term "relativity" as denoting the relative space and time coordinates for two observers who are in relative *motion*. The term, motion, in turn, relates to a *continuous change*. The types of equations that must be utilized to describe such change—differential equations—were first discovered in the seventeenth century by Leibniz and Newton. It was Newton who first used this type of mathematical equation to describe motion. It is interesting to note that in contrast with his description of matter as an assembly of independent, discrete things, it is the *continuous relation* that is basic in the relativistic approach to physics. Here, the relation plays the active role as a fundamental entity, while the "particles of matter" play the passive role of a derived prediction. In contrast, the "particles of matter" play the active role in the description of the quantum theory, while the relation between the particles (that is, the forces that they exert on each other) play the passive role.

When one fully uses relativity in the construction of the mathematical language that is to be used to make predictions about the properties of nature, he finds that the basic elements of the language that must be used are a set of *continuous field variables*. The full set of the field solutions of the basic equations are, in principle, an exact description of the fundamental relations that follow from the assumption of the principle of relativity (that is, the principle which asserts that the laws of nature must have the same form, independent of the space–time coordinate frame in which they are expressed).

The theory of relativity is then, necessarily, a *field theory*. This is in the same sense with which the field concept was originally introduced by Faraday and described mathematically by Maxwell in electromagnetic theory. In contrast, the quantum theory is necessarily a *particle theory*—but not in the same sense that was meant by Newton. The latter difference, as we have indicated earlier, has to do with the nondeterminism that is inherent in the quantum theory of measurement as compared with the deterministic description in classical physics.

In summary, the quantum theory resembles the classical, Newtonian approach as a theory of discrete particles of matter. It differs from the classical theory in the definition of the values of the properties of matter. In the classical case, these values are predetermined and independent of any method of measurement; in the quantum theory, the values are not predetermined, but rather depend on the method of measurement. The corresponding conceptual features of the theory of relativity are just opposite. The latter theory implies that the basic elements of the language that must be used to describe nature are continuous fields and that the set of such field solutions that describe a physical system must provide a predetermined and complete description that is independent of any measurements that might be made on the system. In this way, the theory of relativity resembles the Newtonian theory. On the other hand, since the theory of relativity requires the basic variables that describe matter to be continuous fields, the consequent predictions are that the values of all properties of a physical system that is described in this way must have a *continuous distribution*. Thus, the theory of relativity differs from the Newtonian theory and the quantum theory in its prediction of continuous, rather than discrete quantities to describe the properties of matter.

We see, then, that the quantum theory and the theory of relativity are dichotomous. In other words, a part of the underlying concepts for one of these theories is incompatible (in the logical sense) with some of the conceptual bases of the other theory. In the first quarter of the twentieth century, both of these theories were shown to be successful in their own domains. That is, where one of these theories worked, the other was not required, and vice-versa. However, since some experiments required the use of both of these theories at the same time, their fusion was necessary. But as long as they remained logically incompatible, they could not be fused while still maintaining all of their respective, conceptual basis elements. Thus, to resolve the problem of being able to explain the experimental results of high energy physics, there were two possibilities. Either one must strictly conform to the conceptual notions of the quantum theory of measurement at the expense of giving up some of the concepts of a strictly relativistic theory, or else one must strictly conform to the relativistic field theory at the expense of giving up some of the conceptual notions of the quantum theory. In the following chapter, we will discuss some of the research that has been done with the former approach—the approach that has been taken by the great majority of present-day physicists. In the succeeding chapter, the latter approach will be discussed in more detail.

# 10.

# Relativistic quantum field theory

If one wishes to maintain the entire conceptual structure of the quantum theory, then it follows: (1) that the position and the momentum of a particle cannot be defined at the same time with arbitrary accuracy, and (2) that the fields which may be associated with these particles also cannot be defined at the same time with arbitrary precision. For example, a moving, electrically charged particle of matter—say, the electron—would have an electric and a magnetic field associated with it, according to Maxwell's equations for electromagnetism. According to the field theory, the electric and magnetic field variables have precise values (that are predetermined) throughout all of space and time. Nevertheless, if one should impose on this description the necessity of incorporating the nondeterministic approach to measurement of the quantum theory, then it would follow that not all of the components of the electric and the magnetic field variables for this charged particle could be determined from measurements, with any desired accuracy, at the same time. That is, it would follow from the quantum postulate that one could no longer claim that the electric and magnetic fields have predetermined values throughout space and time. The accuracy with which they can be specified would depend on the conditions of measurement.

To describe this state of affairs in a mathematical way, one then proceeds to convert the electric and magnetic field variables into "field operators." These, in turn, are allowed to "operate on" the state function of the system (the state function for the electron in this case). Thus, suppose that $f$ is a state function that relates to a particular set of properties of the electron. $E_1$, $E_2$, and $E_3$ are the three components of the electric field, and $H_1$, $H_2$, and $H_3$ are the three components of the magnetic field that describes the force field of the moving electron. If one should measure the electric field variable in the $i$th direction (where $i$ is 1, 2, or 3), then the state function is transformed as follows: $f \rightarrow \widehat{E}_i f$. Here, $\widehat{E}_i$ stands for the operator that represents a measurement of the electric field intensity in the $i$th direction. If a measurement should then be performed to determine the strength of the magnetic field in the $j$th direction at the same time that the electric field in the $i$th direction was determined, then $\widehat{E}_i f \rightarrow \widehat{H}_j \widehat{E}_i f$. It has been asserted above that if the order of measurement described had been inverted, the same result would not necessarily follow. That is to say, $\widehat{E}_i \widehat{H}_j f$ is not the same thing as $\widehat{H}_j \widehat{E}_i f$. In quantum field theory this is the way of expressing the contention that not all of the components of the electric and magnetic field variables are *simultaneously* measurable to arbitrarily precise values.

The preceding conclusion is a consequence of the contended nonmeasurability, to any desired precision, of the simultaneous values for all of the components of the position and momentum of the moving electron expressed earlier in the form

$$(\widehat{xp} - \widehat{px})f \neq 0.$$

This, of course, is because the electric and magnetic *fields* depend on the momentum and position coordinates of the electron. Since the practical application of the preceding requirement on the measured position and momentum coordinates applies only to the microscopic domain, the similar restriction on the measura-

42. *Max Planck (1858–1947)*

bility of the electric and magnetic field components also applies, in practice, only to a sufficiently small spatial domain.

This analysis applies not only to the measurements of the electric and magnetic fields, but it should apply, in accordance with the postulate of the quantum approach, to all other field variables for a microscopic quantity of matter. We have argued earlier that the theory of relativity necessitates the use of continuous field quantities, rather than the discrete particle variables, to describe any physical system that is in accordance with the latter approach. Thus, if one wishes to force the quantum theory to be consistent with the *mathematical requirements* of the theory of relativity, the following has to be done. One would have to force the quantum mechanical equations to have the same form in the space–time frame of reference of any observer who is in relative motion compared with any other. The fields (which are the quantum mechanical wave functions that tell us by their variation how the probability of measuring the particle in a state must change with time) must also be converted into "field operators." Thus, the Schrödinger wave equation itself is converted into a wave operator and applied to the set of state functions for the system of particles that is observed. This mathematical scheme is called *second quantization.*

When one proceeds in this fashion to construct a relativistic quantum field theory, the equations that are obtained resemble in form the standard, quantum mechanical equations. When they are studied in detail, it is soon discovered that these equations do not have any solutions! Nevertheless, certain techniques have been developed (since the late 1940s) which prescribe a set of rules (having to do with a manipulation of these equations) in order to yield numbers that are to be compared with observed properties.

The application of this theory to a description of the motion of electrically charged particles that interact with each other in accordance with the electromagnetic theory is called *quantum electrodynamics* (or *qed,* for short). The application of *qed* to actual physical situations has thus far given two predictions (that are in remarkable agreement with experimental observations) of effects that are not predicted at all by the usual, or first, quantized theory. One of the predictions has to do with some extra energy levels of the hydrogen atom that were not anticipated by the ordinary wave equation. Both the existence of these extra energy levels and their quantitative separation have been predicted by *qed* and are in numerical agreement with the data to one part in ten thousand! The latter effect is called the *Lamb shift*—named after W. E. Lamb, Jr. (1913–    ), who directed the experimental investigations that led to these observations. The second prediction that was indicated above has to do with the magnetic properties of the electron. The numerical prediction for the "magnetic moment" of the electron that was obtained from qed agreed to within one part in ten thousand with the experimental result. This prediction is much closer to the data than is the prediction of the ordinary quantum mechanical theory.

The extremely close agreement of these two predictions of qed with the experimental observations greatly encouraged physicists from the late 1940s on and inspired confidence in the scheme of computation that was being used. This was true even though the existence of the solutions of the actual equations of quantum field theory could not be established. It was hoped that the combination of the equations of quantum field theory, together with the scheme of calculation that was invented, might, in fact, represent a first approximation to the actual solutions of the equations of quantum field theory, if the latter theory could be expressed in its proper form, that is, the form that would exhibit the existence of solutions. Although the latter expression of quantum field theory has not yet been found (at the time of this writing), the overwhelming confidence that the present physics community does have in the quantum approach makes it imperative for us to look, in more detail, at the physical interpretation that is now being given to the formal expression of quantum field theory, in its attempt to describe the high-energy phenomena that are presently being observed in laboratories all over the world. One of the essential concepts in this approach, that we will now discuss in more detail, is that of creation and annihilation of matter.

### Creation and Annihilation of Matter

In addition to the predictions of quantum field theory that have been discussed, one can also associate observable events with the field operators themselves. Recall that a field operator is a continuous field variable of the relativistic theory that has been converted into a linear operator—an instruction that must be given to a wave function. Since the original, continuous function is defined at each point in space–time, the field operator, into which this has been converted, must correspond to an "instruction" at each point in space–time. Now, since the Schrödinger wave function is supposed to relate the probability for locating a particle at one place or another in space, how should its corresponding field operator be interpreted? The interpretation that is given is that the "field operator" form of a Schrödinger wave function should represent the "creation" of a particle (at the space–time point where it is defined). But, before the particle was "created," there was no particle! Thus, one of the state functions, upon which these field operators act, corresponds to a universe without any matter in it. We call the state of no particles of matter or radiation the *vacuum state*. When the field operator which relates to the creation of a particle acts on the vacuum state, it creates a new state—the state of one particle. Similarly, when the field operator describing the creation of a particle acts on a one-particle state, it creates a two-particle state, and so on. When the "creation operator" acts $M$ times on an $N$-particle state, it creates an $(M + N)$ particle state.

Similarly, one can define a Schrödinger field operator to "annihilate" a particle. The "annihilation operator" is called the "conjugate" of the "creation operator." When such an operator acts on a one-particle state it generates the vacuum state. When it acts on a two-particle state, it generates a one-particle state, and so on. When the annihilation operator acts $M$ times on an $N$-particle state, it generates a state that describes $(N - M)$ particles.

In the foregoing description, we see that the significant quantities that characterize the state function are the *numbers* of particles of matter that are present in the system rather than the energy or momentum of these particles. The actual derivation of these numbers follows, in this theory, from average values of the commutators or anticommutators of the different field operators. The commutator of $A$ and $B$ is $(AB - BA)$, the anticommutator is $(AB + BA)$.

To apply quantum field theory to moving, electrically charged particles (*qed*), we first convert the field variables that appear in Maxwell's equations for electromagnetism into field operators. Next, we combine these with the "matter field operators" to form an operator that represents the interaction between the electromagnetic field and the matter that is in motion. (The "matter field operators" are the Schrödinger-type wave functions that have been converted into operators in accordance with our previous discussion.) The resulting "interaction operator" then acts on one of the state functions for the physical system. When the operators are defined as "creation" and "annihilation" operators, the electromagnetic interaction operator is found to contain terms that must entail operators describing the annihilation of a pair of oppositely charged, and equally massive particles (which we will call "electron" and "positron"), and at the same time, the electromagnetic interaction operator must also relate to the creation of a pair of photons. According to this description, the two photons must be polarized in a common plane, having a 90° phase difference, and proceed in opposite directions along the axis that is perpendicular to their common plane of polarization. In addition to the preceding "prediction," the interaction operator also contains a term that corresponds to the opposite reaction, in which an electron–positron pair is created and the photons are simultaneously annihilated.

The preceding discussion relates to an *interpretation* of the *equations* of quantum field theory when they are applied to a description of moving, electrically charged particles. But the question should then be asked: Are there really such events where matter is annihilated and created? In experimental high energy physics, certain events are indeed observed that can be identified with this type of reaction. For instance, one sees in a cloud

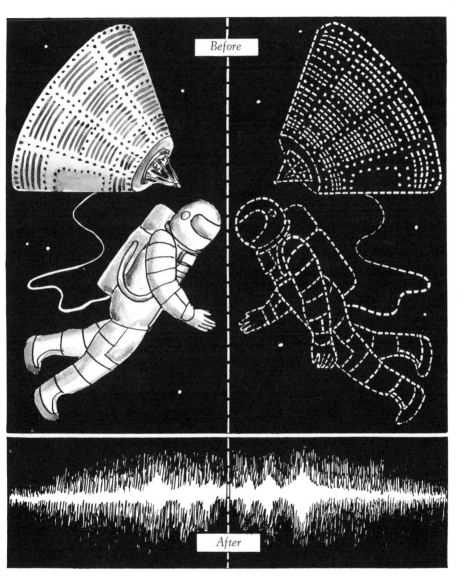

43. *The Annihilation of Matter-Anti-Matter.* (a) *Before: John Jones approaches anti-John Jones.* (b) *After: They annihilate each other's individuality, converting their intrinsic energies into radiation which then propagates away at the speed of light.*

chamber (or in a bubble chamber) that a pair of tracks of equal thickness approach each other and join at a vertex; then no more tracks are seen to be connected. Cloud chambers and bubble chambers are devices used for detecting charged particles. Droplets or bubbles are formed on the ion trails left as the particles pass through a liquid or vapor. In the cloud chamber, the vapor is supersaturated by a sudden temperature change, and in the bubble chamber a sudden reduction in pressure forces the liquid over its boiling point. Further experimenting has established that in coincidence with the joining of these tracks at the vertex, one can detect two electromagnetic currents, whose phases are correlated with a 90° difference and are polarized in a common plane. The pair of initial tracks, were then identified with the electron (matter) and the positron (called antimatter). The final, correlated currents were identified with the two "photons" that are supposedly created when the particle–antiparticle pair is said to be annihilated at the vertex.

Thus, there is indeed experimental evidence that would support the interpretation of the equations of quantum field theory which provides for (1) the creation and annihilation of matter, and of quanta of the electromagnetic field, and (2) a symmetry between matter and antimatter. But this is the first time that we have even mentioned the existence of the *positron*— the antimatter that must annihilate simultaneously with the matter of the electron field. Where does the positron actually originate in the theory? Are there any stronger reasons for the necessity of the existence of a symmetry between matter (electrons) and antimatter (positrons)? In the next section we will try to answer this question.

### The Dirac Equation

Before the rules of second quantization were applied, P. A. M. Dirac showed how the Schrödinger form of the quan-

tum mechanical equations could be expressed in a way that satis-
fies the *mathematical requirement* of the theory of relativity.
The resulting Dirac equation maintains its form in the space–time'
coordinate system of any observer who was in motion relative
to any other observer. The Dirac equation takes the Schrödinger
form exactly, in the limit where the relative speeds between
observers are sufficiently small compared with the speed of light.
Thus, in this "nonrelativistic limit," everything that is predicted
by the Schrödinger wave equation is also predicted by the Dirac
theory. But in the exact form of the Dirac equation, extra predictions are made. One of these has to do with the energy level
spectrum of an excited gas that is observed in a magnetic field.
In these experimental observations, *extra* energy levels were seen
that were not at all explained by Schrödinger's equation. However they did fit exactly into the predictions of Dirac's equation.
Thus, Dirac's equation stood as a generalization of the Schrödinger wave equation for quantum mechanics.

The essential feature of the Dirac equation that made it
supersede the Schrödinger equation was that its solutions depended on extra variables in addition to the space and time
coordinates. These extra coordinates, called "spin," were, in fact,
necessitated by forcing the form of the original Schrödinger
equation into a relativistically covariant expression. The spin
variables implied that the angular momentum of an electron is
in two parts. One is the usual angular momentum that depends
on the spatial coordinates of the particle, relative to the point
about which it is rotating. The second part, however, is independent of any spatial coordinates. This is the portion of angular
momentum of the electron that depends on the spin variables.
It is called "intrinsic angular momentum." Thus, it was discovered that even if the electron would not be rotating about
some origin in space (say the nucleus of an atom), it would still
have some angular momentum left.

The "electron spin" was actually postulated before Dirac's
discovery, by G. E. Uhlenbeck (1900–    ) and S. A. Goudsmit

(1902–    ), in order to explain the observed spectrum of radiating atoms in an external magnetic field. The mathematical insertion of these "spin variables" into the Schrödinger equation was carried out correctly by W. Pauli (1900–1958). Nevertheless, the actual origin of electron spin was not discovered until Dirac derived these extra degrees of freedom for the electron, as a consequence of forcing the Schrödinger equation to also satisfy the mathematical requirements of the theory of special relativity. The solution of Dirac's equation was found to be a four-component entity, called a "spinor," rather than the single component function that solved the original Schrödinger wave equation. However, as we have indicated earlier, the nonrelativistic limit of the Dirac "spinor" solution is precisely the single-component solution of the Schrödinger equation.

In addition to predicting correctly the physical effects that could only be attributed to electron spin, the Dirac equation also predicted that for every particle that exists, there must also be an antiparticle. The antiparticle has the same mass and same intrinsic angular momentum as the original particle. It carries a charge that is equal in magnitude, but with opposite polarity, to that of the electron. Since all known matter in the universe is composed of elementary particles that must include electrically charged components, Dirac's discovery implied that the universe must be composed of a distribution of matter and antimatter.

It followed from the Dirac theory, as well as from its successor, the quantum theory of fields, that when matter and antimatter interact they will annihilate each other, thus creating electromagnetic radiation. To conserve energy, it is clear that if a particle of mass $m$ and its antiparticle (which also has a mass equal to $m$) should annihilate each other, their combined rest energy, which according to Einstein's analysis is equal to $2mc^2$, must be the minimum energy of the electromagnetic radiation that is thereby created.

In view of the fact that all of our direct experience is in

*44. Wolfgang Pauli (1900–1958)*

terms of the transfer of energy and momentum between quantities of inertial matter, one would have to conclude, according to the Dirac theory, that there is much more matter than there is antimatter in our particular region of the universe. For if the abundance of matter and antimatter were the same in a given region of space, there would be complete annihilation of all existing matter. Nothing would be left but electromagnetic radiation! Thus, the fact that there is matter here means that an imbalance between matter and antimatter must exist in this region. It is, of course, only convention that makes us call the material in our region "matter" and not "antimatter." However, there is no reason why there should not be other regions of the universe in which it is antimatter, rather than matter, that predominates in abundance.

The preceding conclusions of the Dirac theory lead to some interesting speculation. It is conceivable, in accordance with this view, that somewhere in the universe there is an anti-earth. This is a planet that is exactly the mirror of ours, except that all matter is replaced with antimatter. Suppose that we carry the speculation one step further by assuming that for every John Jones on earth, there is a counterpart on anti-earth, let us call him anti-John Jones. The gentleman, anti-John Jones, is a man who corresponds precisely to John Jones, in the sense that every atom of matter that is included in the man, John Jones, corresponds to an antiatom of matter in the man, anti-John Jones. With such a situation existing in the world, it would be best if these two men did not meet each other. For if their meeting could be arranged, John and anti-John would automatically annihilate each other. They would lose their individuality as they meet and they would transform into electromagnetic radiation (Figure 43).

Of course, such a test of the validity of the matter-antimatter annihilation process is not a satisfactory one for several reasons! A more practical approach is to look for some antimatter here on earth, that is, to look into a cloud chamber for the tracks of particles and their counterpart antiparticles when they are supposedly being "annihilated" or "created" in high-energy processes. As we have argued earlier, the full use of the quantum approach to measurement theory would necessitate the formalism of quantum field theory. In the latter description, the Dirac spinor solutions have to be re-expressed as "field operators." Thus, these field operators, in the second quantized theory, must represent the creation and annihilation of particles of matter that would have the extra angular momentum that was associated with the spin coordinates in the first quantized Dirac theory. The latter extension from the Dirac theory to the quantum field theory was logically necessitated by the conceptual basis of the quantum approach in a form that is compatible with the mathematical requirement of the theory of relativity.

Aside from the logical requirement for extending the Dirac theory to a relativistic quantum field theory, there were two outstanding deficiencies in the predictions of the older theory.

One of these had to do with the energy level spectrum of hydrogen. While the Dirac equation did give a much better prediction of this spectrum than the Schrödinger equation, it was still unable to predict the extra energy levels that were observed by Lamb. As we indicated before, this problem appears to be resolved when the first quantized theory is extended to quantum electrodynamics. However, we must still ask the question: What are the *physical assumptions* that are used in quantum field theory that are not present in the older Dirac theory, that lead to a *mechanism* that gives rise to the extra observed energy levels? Let us now try to answer this question in terms of some of the essential differences between the two expressions of quantum mechanics.

We have seen earlier that quantum field theory is a many-particle theory, while the Dirac theory of hydrogen has to do only with the binding of one electron and one proton, *and nothing else*. On the other hand, the model of the hydrogen atom that emerges from the formalism of quantum field theory *necessarily* pictures the bound electron–proton system inside of a "sea" that is made up of an indefinitely large number of electron–positron pairs that are in a continual state of transmutation into electromagnetic radiation and vice-versa. In hydrogen, the electron that is bound to the proton must also be a part of this sea of pairs of charged particles and radiation. The latter background for all matter, according to quantum field theory, is the so-called *physical vacuum* state. Thus we see that this sort of "vacuum" must have physical consequences. It is not a "vacuum" in the sense of "nothingness." One of the consequences of the physical vacuum is to shift the energy levels that correspond to certain Dirac solutions so that pairs of solutions which originally corresponded to the same value of energy now correspond to different, distinguishable energy values. This comes about because of the electromagnetic coupling between the electron that is bound in the single hydrogen atom and the infinite sea of radiation and elec-

*45. Paul Adrien Maurice Dirac (1902–     )*

tron–positron pairs that constitute the "vacuum" in which that atom is embedded. The quantitative value for this Lamb shift in the different states of hydrogen that has been measured is in remarkable agreement with the predictions of quantum field theory. It is not at all predicted by the older Dirac theory (which in turn is a relativistic generalization of the Schrödinger wave equation for quantum mechanics). This prediction of quantum electrodynamics, which was not made by the Dirac theory, gave the community of physicists a great deal of confidence in the theoretical reasoning that originally necessitated a conversion from the ordinary quantum mechanics to quantum field theory.

It is also because of the coupling of a free electron to the "physical vacuum" in quantum field theory that the prediction of its magnetic moment in the latter theory is different from the predicted value in the ordinary Dirac theory. The experimental data confirms the prediction of quantum field theory to quite high accuracy.

Another outstanding feature of the *observed* properties of atoms, not in agreement with the predictions of the Dirac equation, is the *lack* of certain transitions between atomic energy levels that are implied by the latter equation. These have to do with the existence, in the Dirac theory, of *negative* energy levels. Now, peculiar as it may seem to those who might think of "energy" in the conventional terms, there is really nothing that would rule out the existence of negative energy. This is because it is the *energy differences* that are actually identified with the experimental observations. In classical physics, the concept of energy is intimately connected with *continuous change*. Thus, if all of the possible values of the energy of a physical system were negative, or if all were positive, one could always make a change from one energy value to another in a continuous fashion.

On the other hand, if some of the energy values were positive, and others negative, no transition could be made from the positive energy values to the negative energy values in a continuous fashion. The change in sign of the energy could be compared, for example, with the change in sign of the direction of a light beam when it is *reflected* from a mirror. Thus, there cannot be both positive and negative energy values available to a physical system, according to classical Newtonian mechanics, or in a continuum field theory, such as the one that is implied in the theory of relativity. On the other hand, the changes in energy in quantum mechanics occur in a *discontinuous fashion*. Thus, there is nothing to rule out a transition between a state of positive energy for an electron in an atom, say, and another one of its states of negative energy. Thus, the Dirac theory predicts that if two hydrogen atoms with sufficient energy collide, one atom can transfer a quantity of energy to the other that is equal to the difference between the positive and negative energy values. Thus, one might expect to see energy transfers in such collision processes that are of the order of $2mc^2$, where $m$ is the mass of the electron. Such quantities of energy transfer are not in fact seen in atomic scattering experiments. Thus, the experimental facts appear to rule out the existence of negative energy levels, thereby refuting the predictions of the Dirac equation. A change from the Dirac to the quantum field theory to describe the hydrogen atom was required for the following reasons: the observation of the Lamb shift and the lack of observable proof of the effect of negative energy levels.

### Difficulties in Quantum Field Theory

When the relativistic form of quantum mechanics—the mechanics embodied in the Dirac equation—is "second quantized," the newly formed mathematical equations of quantum field theory are a great deal more complicated than the earlier ones. For one thing, one exchanges a description of matter in terms of one or two interacting bodies at a time, with an infinite number of bodies of matter and radiation that are in continual interaction. In electrodynamics the latter is the infinite sea of electron–positron pairs and photons that are in a continual state of creation and annihilation—forming the "physical vacuum" into which we must insert the atom before we can determine its observable properties. At first sight, one might think that perhaps nature is not really that complicated! But then, the theory might be acceptable, after all, since it does indeed make some successful predictions that are not made by the other (simpler) expressions of quantum mechanics.

On the other hand, the mathematical form of quantum field theory is still not in an acceptable state, for several reasons. First, because of the nature of the physical model that is evoked here, the predictions of all of the physical properties of matter are expressed only in terms of infinite sums of numbers. The main difficulty here has to do with the question of what it is that these sums of numbers add up to. As a matter of fact, before the applicability of certain "subtractions" were discovered, these sums always added up to infinity! This implies that the original formulation of the theory must have been wrong since these sums are supposed to represent the *finite values* for corresponding, observable properties of matter. It was the purpose of the "renormalization method" in quantum field theory, to *subtract out* the infinite terms of the original sums, in order to yield the finite numbers that are required to compare with the actual experimental facts. As indicated earlier, this mathematical scheme led to the predictions of two experimental results. One, the Lamb shift in the energy levels of hydrogen-like atoms, and two, the magnetic moment of the electron. Both results were in remarkable agreement with the measurements. Nevertheless, it is not yet clear that without changing any of the physical description of the hydrogen atom, equivalent mathematical schemes might not also predict any other value that we please for the Lamb shift. This is another way of saying that the "renormalization technique" in quantum electrodynamics cannot be shown to be mathematically consistent.

As we have discussed earlier, it is essential to the method of theoretical science that any scheme for predicting the properties of nature must necessarily be both logically and mathematically consistent. If we adhere to this requirement, then it must be said that the present scheme of prediction that is incorporated in relativistic quantum field theory and quantum electrodynamics has not yet established itself as a bona fide physical theory. This difficulty is serious indeed, not only in the description of relativistic quantum mechanics, but also in regard to nonrelativistic quantum mechanics. If the ordinary, nonrelativistic quantum mechanics is supposed to be an approximation to a relativistic theory of quantum mechanics and if the latter theory is relativistic quantum field theory, then it must follow that in the eventuality that the latter theory cannot be proven to be a bona fide theory, obviously neither can any of its approximations be considered bona fide.

On the other hand, should one be willing to give up the requirement of relativistic invariance altogether, then nonrelativistic quantum mechanics would remain mathematically consistent, and within Bohr's interpretation, logically consistent. It would therefore qualify as a bona fide theory to describe microscopic matter when having sufficiently low energy. However, it is not possible to give up the requirement of relativistic invariance because we know that the observed low-energy events in the atomic domain are indeed limiting properties of observed high-energy events. That is, by starting with an experiment in high energy physics and continually decreasing the energy transfer between the microscopic quantities of matter that are involved, we see how the observed high-energy effects approach the observed low-energy effects.

Thus, the experimental observations demand that the theoretical description of microscopic matter must unify the non-relativistic with the relativistic domains (high energy with low energy). Relativistic quantum field theory has attempted this unification by incorporating the mathematical requirement of the theory of relativity within a theory that makes full use of the conceptual basis of the quantum theory of measurement. The form of the equations must be unchanged in transformations between the space–time coordinate "languages" of any two observers. This attempt has, unfortunately, not yet succeeded.

The failure to provide a demonstrably consistent, relativistic quantum field theory has led to two separate alternative steps to resolve the difficulty. The first is to alter the mathematical structure of the equations, while still maintaining full use of the ideas that underlie the quantum approach (discreteness, non-determinism, etc. ). The second alternative would be to take the

Just as in the case of quantum field theory, nuclear physics also does not have a set of mathematically consistent field equations that would completely describe the way in which nuclear particles of matter affect other nuclear particles. For example, the motion of protons in the "field" of other protons or neutrons is not specified here as it is in the case of electrodynamics where the motion of charged matter under the influence of other charged matter can be predicted. Of course, a great deal has been learned about the nature of nuclear forces, such as their short range character. Unlike the electromagnetic or gravitational forces, which extend themselves over all of space, the nuclear force (which is very much stronger than the latter types of forces in the range where it is effective) approaches zero when the distance between nuclear particles becomes greater than the order of $10^{-13}$ cm. It is, in fact, the short range feature of these forces that have hindered our attainment of understanding more of nuclear physics.

A very clever scheme was proposed in the late 1930s by J. A. Wheeler (1911–    ) to analyze nuclear forces without having to know the specific field equations that give rise to these forces. This analysis considers the possible final state functions that would relate to the nuclear particle that had scattered from another nuclear particle, in terms of the state function before the scattering had taken place. Since the nuclear force has a short range, the projectile nuclear particle does not "know" about the existence of the target nuclear particle until it comes very close to it. Thus, before the scattering has taken place, the projectile nuclear particle can be considered as a "free" particle. The state function for the particle, before it scatters, is then the mathematical expression for a free wave that is propagating in a particular direction (according to the definition of this function as a solution of a "wave equation"). This wave function relates to the probability for measuring the location of the particle in one place or another (wherever the wave function is evaluated).

Similarly, the short range character of the nuclear force implies that very soon after scattering has taken place (by means of the nuclear force), the scattered particle once again has the appearance of a free particle. The latter state function has the same sort of "wave" behavior as its initial state function. The only difference is that the phase of the wave has been shifted by the scattering process.

To exemplify the latter description of the scattering of waves by a short range force, consider the ocean waves that scatter from a buoy that is fairly stationary in a particular location. The waves of water that approach the buoy look just like the waves that scatter from it, especially when one looks at these waves not in the immediate vicinity of the buoy. The difference in the oncoming and the scattered ocean waves is only in their phases. It is clear that there is a phase difference in the two waves because of interference that the scattered wave has with other portions of the oncoming waves.

The amount of phase shift in our nuclear scattering problem clearly depends on the strength and duration of the scattering force. The analysis of the scattered nuclear particles, in an actual experiment, leads to a prediction of the phase shifts which, in turn, are used to analyze the features of the nuclear force itself. To sum up then, the nuclear forces are probed by studying the different quantum mechanical state functions that relate to the nuclear particles, *long after* and *long before* the scattering event has taken place. (The latter functions are called "asymptotic solutions.")

Just as quantum field theory introduces "operators" that describe the creation and annihilation of particles of matter, so an operator is also introduced in the scattering problem to relate the state function for the "free," scattered particle to the state function for the "free," unscattered particle. The latter is the S-operator, and is conveniently represented in terms of a matrix, called the S-matrix. If the S-matrix is known for a given nuclear scattering process, then it must contain the details of the nuclear forces or at least the details that cause the nuclear particle to scatter in a particular way. Even if we do not know the detailed nature of the nuclear forces, it has been found, on the basis of certain very general considerations, that the S-matrix must have certain specific properties. Without going into the derivations of these properties, let us now briefly discuss some of them.

An important property of the S-matrix that follows from the interpretation of the quantum mechanical state functions in terms of probabilities is "unitarity." This is the name given to a matrix that has a certain symmetry relating its elements. A second important property of the S-matrix follows from the "linear superposition principle" of the quantum theory.

According to the definition of the S-matrix, this principle, which is a necessary requirement of the quantum theory, implies that the elements that make up this matrix must be independent of the possible incoming and scattered wave solutions. A third important property of the S-matrix is implied by the assumption of "time reversal invariance." This refers to the assumption that nuclear forces do not distinguish between forward and backward time directions. This means that if time is reversed, all of the scattered particles would precisely retrace their earlier paths.

The attempt is then made to derive the S-matrix. The S-matrix is structured to have the properties that are implied by the general conditions mentioned above. After the general form of the S-matrix is determined, it is then applied to the nuclear scattering problem, where the phase shifts that appear in this matrix are identified with the observed effects. Finally, the phase shifts, so-determined, are used to deduce some of the detailed properties of the nuclear potential energy that caused the scattering to take place.

This procedure in the study of nuclear forces might be compared with a blindfolded man, throwing a very large number of tennis balls at a barn. Some of these balls would miss the barn, some would bounce back off of the wall, some would go through a window and stay inside the barn, and still others might go in one window and out another. The game is then to challenge the man, who is blindfolded, to count the tennis balls that have landed on the ground and, from their spatial distribution, to deduce the nature of the barn—without ever being allowed to look directly at it.

The use of the scattering matrix was very effective indeed in drawing out some of the important features of nuclear forces that we know about today. Thus, this method was considered, in the 1950s, as a possible way to describe high energy physics, without having to consider the actual equations of relativistic quantum field theory. The direct examination of the latter equations would be analogous to looking directly at the barn. Thus, from research in the 1950s and 1960s, it was found that additional predictions could indeed be made in regard to the scattering of highly energetic particles. Nevertheless, the more ambitious proponents of this approach have the hope that the S-*matrix theory* will eventually explain all of the physical processes in the relativistic domain of elementary particle physics. Still others cannot agree with this view. There are several reasons for the opposition. First, the S-matrix approach has not yet given a satisfactory account of the long range forces (electromagnetic and gravitational forces). It would seem strange indeed that this could claim to be a *general law* and yet only be compatible with some types of natural forces and not others.

A second difficulty with the S-matrix approach is that *everything* seems to be described in terms of the scattering of matter from matter. But what about those physical processes that involve matter that is *bound* to other matter—where the interacting parts of a system can never be described as free, or almost free? Opponents would argue: Does it seem meaningful to describe the orbital motion of the earth in its planetary motion under the influence of the sun, in terms of the *scattering* of the earth from one "free" state to another? If one should insist on such a description of the bound state (nonfree), then the model that is used must entail an *infinite* number of scatterings in which the earth is thrown off of its orbit and then returned again in each of these events. This would have to happen in such a way that *on the average,* the earth would always be *observed* where we know the orbital path to be.

Finally, there remains discomfort among those who believe that a completely ordered set of field equations must describe the properties of nature, if they are confronted with the conjecture that it is not meaningful to ask for such a set of equations. For even if the S-matrix approach were entirely successful in predicting all of the known properties of nuclear forces, and even if it succeeded in the future to make a great deal of correct predictions about high energy physics, in general, it still does not seem to be compatible with the nature of other phenomena, such as electromagnetic interaction and gravitation. Even if it could be made compatible with the latter phenomena, these phenomena already have a perfectly well-behaved set of underlying field equations that do give extremely accurate results in their own domains of prediction. The question then follows: Why should it be *meaningful* to say that field equations underlie some physical phenomena yet not meaningful to expect field equations to underlie other phenomena?

In spite of these objections to the S-matrix approach to a theory of matter, it is still true that the net effect of the great amount of effort that has been given to the mathematical description of high energy scattering events has been positive. It has indeed enhanced our understanding of at least the mathematical description of the whole class of observations of interacting matter identified with the scattering process.

A survey and reprinting of some of the original works on significant mathematical developments in S-matrix theory is given by G. F. Chew in S-MATRIX THEORY AND STRONG INTERACTIONS (Benjamin, 1961).

## Axiomatic Field Theory

The S-matrix approach favors the complete abandonment of the equations of relativistic quantum field theory, in order to reach an accord between the concepts of the quantum and relativity theories. *Axiomatic field theory* advocates that compatibility might be reached only by widening the axiomatic basis of the quantum theory without abandoning the field equations.

One axiom that would be reasonable to add to the postulates of the quantum theory is the assertion that the vacuum state in quantum field theory, corresponding to zero energy, is the state of minimum energy. This axiom would then resolve the difficulties encountered when negative energy values are predicted by the mathematical equations. It would say that such states of the system must simply be excluded from the set of all possible states.

One other axiom that must accompany this axiom, within the description of quantum field theory, is the assertion that in the complete set of field operators used to describe a physical system, there cannot be one that will "annihilate" the vacuum state.

Of course, the axiom of the principle of relativity heads the list; however, this is necessarily restricted to the principle of special relativity in axiomatic field theory. (Recall that this has to do with relative frames of reference that are at rest or moving at *constant* velocity with respect to each other.) Beside this principle and the assertions about the "vacuum" state, as axioms for quantum field theory, the axiomatic approach adopts several extra postulates that are more of a mathematical nature. For example, it is postulated that the complete set of state functions that describe the measurable properties of matter are infinite in number and they are relatively oriented so that a measurement involving any one of these states will "project" a linear sum of all of them. Such a set of "oriented" state functions forms a linear vector space. In addition to this assumption, it is postulated that the infinite sum of the squares of these oriented state functions is a constant positive number for all values of the space–time coordinates in the same way the sum of the squares of the perpendicular sides of a right triangle is always equal to the square of the hypotenuse of the triangle.

Another axiom that is adopted is that all interactions must be local. This means that matter at some distant point cannot affect matter that is at some different point in space if the former field of matter has no value at the location of the latter quantity of matter. The assumption of "locality" then implies within the theory that the field operators that relate to any measurement must either commute with each other or they must anticommute with each other. By commutation we mean that the order in which the operators appear in an expression is irrelevant so that $\widehat{ab}$ will give the same result as $\widehat{ba}$. If a reversed product changes sign, that is, if $\widehat{ab} = -\widehat{ba}$, it is called anticommutation.

Several of the features of the phenomena observed in the behavior of microscopic matter at high energies can be described by using (1) the postulates of the relativity principle, (2) the properties of the vacuum that were asserted above, (3) the assumption about locality of interactions, and (4) some other axioms. The aim, then, is to use this widened axiomatic base of quantum field theory as a starting point from which to proceed in deriving further physical properties of matter.

Unfortunately, very little has emerged from this approach in regard to actual predictions. Nevertheless, this should not be too bothersome. For the primary aim of the axiomatic approach is to provide a consistent, mathematical scheme in which to incorporate the present form of quantum field theory. Since the latter scheme has already given some correct answers about the properties of nature, such an incorporation would automatically yield a theory that would have the same successes. Thus it appears that the approach of axiomatic field theory is a most important one since its aim is to make a bona fide theory, that is, a consistent set of rules of prediction for the high-energy description of quantum phenomena. If it turns out that axiomatic field theory will succeed in recasting quantum field theory in a mathematically consistent form, it will have accomplished the task of giving the entire quantum approach more credence as a possible general theory. Until it does so, however, the validity of the quantum approach to fundamental processes remains on shaky ground in spite of its successes in describing microscopic matter where relativity theory need not be used! If it succeeds in its task of providing a mathematically consistent theory, it will have gone a long way in supporting the concepts of discreteness and nondeterminism in the fundamental description of matter. If it does not succeed, these underlying notions of the quantum theory will remain in a purely speculative state without the right to claim themselves as "established" knowledge.

As an empirical system of formulas that correspond to the actual data, the application of the S-matrix theory in nuclear physics has given some correct, quantitative numerical results. To be successful, then, axiomatic field theory should demonstrate how the S-matrix theory fits into the predictions of the actual field equations. Of course, axiomatic field theory should also show how the ordinary Dirac and Schrödinger equations of quantum mechanics must follow as a limit of the equations of relativistic quantum field theory. This has not yet been done. Until this is accomplished, it will not be clear that the quantum mechanics of Dirac and Schrödinger, and quantum field theory, are mathematically connected, even though they do have a common basis of ideas.

A technical summary of the results of "axiomatic field theory" is given in AXIOMATIC FIELD THEORY, Vol. I, ed. by Chretien and Deser (Gordon and Breach, 1966).

## Coming Back to the Creation and
## Annihilation of Matter

While it is possible to have a consistent theory, and to have such a theory derived from a form of quantum theory, these criteria would not be sufficient for someone who demanded a complete theory. For such a person, there is one set of experimental facts that he would insist be derived from the true laws of nature. These are the data that are conventionally interpreted as the creation and annihilation of pairs of particles and antiparticles. In quantum field theory, one merely states that this happens. A language is invented that involves creation and annihilation operators, and state functions, that depend on the number of particles and antiparticles under consideration. The creation (or annihilation) of matter is represented by the action of these operators on the state functions of the system. Suppose, however, that one should ask the question: What is the actual physical mechanism for annihilating or creating matter? Other questions in this regard are the following: When a particle and an antiparticle meet each other, where and when do they annihilate each other? When a pair of particles of matter are created from a vacuum, precisely where, when, and under what exact circumstances, does this happen? The quantum theory of fields provides no answers for these questions. For, according to the conceptual basis of this theory, such questions are meaningless. This theory must be content to merely say that these things happen. It is for the same reason that the quantum theory cannot ask precisely when, where, and why, a particular apparatus picks a particular state of the observed microscopic system, out of many possible states. The fundamental nondeterminism of the quantum theory of measurement denies that an exact description exists that could answer these questions with precision. Yet, one does see a pair of tracks in a cloud chamber that do come together toward the vertex of a V—and then nothing! The advocate of the complete theory is then challenged to predict these same observations from the solutions of his equations if the latter equations claim to represent reality. This question will be discussed further in the following chapter, where investigations of an alternative approach to unifying the quantum and relativity theories will be discussed. The latter approach is the one in which the notions of a deterministic relativistic field theory is adopted, fully, at the expense of giving up some of the basic concepts of the quantum theory of matter.

## On Relative Simultaneity in the Quantum Theory of Measurement

A very important fundamental difficulty that persists in the relativistic quantum field theory that would have to be removed by the enlargement of the axiomatic basis of the theory is the following: When one asserts that two things happen at the same time, the theory of relativity would require the specification of the frame of reference in which the happening takes place. Since different space–time frames of reference specify different time coordinates, two events that may be simultaneous in one frame of reference would not necessarily be simultaneous in another. The idea of *simultaneity* is a relative concept in relativity theory, while it is independent of the frame of reference in the classical theories of Newton and Galileo. On the other hand, the quantum theory separates the measuring apparatus from the observed "microscopic matter." It asserts that the measured numbers that relate to the interaction between the matter and the apparatus must drop out, simultaneously, from the observation of all *relatively moving* components of the system of matter. In this description, one cannot take the measuring apparatus to a different inertial frame of reference where it could then "see" two separate time values that correspond to the previous simultaneous events. But this conclusion of the quantum theory seems to destroy the notion of *relative simultaneity* that is so deeply embedded in the relativistic approach. Thus, there appears, at this point, to be a fundamental dichotomy between some very basic notions that underlie the quantum and relativity approaches to theories of matter. At the present stage, it is difficult to see how axiomatic field theory might resolve this logical inconsistency from a mathematical model.

It could be conjectured, now, that even if the axiomatic field theory should succeed in resolving all of the present difficulties of the quantum theory, prejudice would still persist within the scientific community against it. This is because of the feeling among many that the true laws of nature should rest on one or two axioms and that these axioms should be strongly hinted at by our *physical experiences* with the world. Thus, Einstein attempted to construct a general theory that would rest essentially on one axiom—*the principle of relativity*—an axiom that can certainly be understood (with some thinking!) from experience. Einstein, and the others who take his view, would most probably have found it difficult to believe that it should be necessary to introduce a dozen or so *mathematical axioms,* very few of which seem to be experienced physically, in order to provide a true law of nature. This attitude is best summarized in Einstein's famous comment: "God may be subtle, but He is not malicious." On the other hand, should axiomatic field theory eventually succeed, a proponent of the approach of quantum field theory may rebut: It is not that God is malicious, it is rather that we simple-minded humans have not yet been able to learn His language sufficiently well to express His laws in a simple form. Of course, these are views that one scientist may have and another may not. They do, however, play an important part in the initial choice of one approach or another to a general theory.

## The Recent Dirac Theory

During the 1960s, one of the creators of the quantum approach to fundamental processes, P. A. M. Dirac, has introduced modifications into the standard form of the quantum field theory, in an attempt to provide a resolution of this forty year old problem of unifying the quantum theory with the theory of relativity, in a mathematically consistent way. In his recent studies, Dirac has assumed that the Schrödinger representation, in which one describes the time development of the state function, must be abandoned altogether. He has shown that, in the realm of quantum field theory, the Schrödinger and the Heisenberg representations are not really equivalent. Dirac then chooses to maintain the Heisenberg representation as the more fundamental one. (Recall that this is the description in which one considers the time development of the *operators*, rather than the *state functions*, which in turn relate to the measurement itself.) In this new theory, Dirac takes the radical step which maintains that these operators, which relate to the act of measurement, *are the observables*.

Dirac found that when this view was pursued to its logical extreme, it no longer was possible to interpret the state function in terms of the probability for measuring the particle in one state or another. Dirac's recent theory does provide a formalism which corresponds to the ordinary quantum electrodynamics in regard to its predictions about electromagnetic phenomena. This theory, then, correctly predicts the energy levels of hydrogen atoms, including the Lamb shift; it also correctly predicts the magnetic moment of the electron. Unfortunately, however, it has not yet reached a state where the infinite quantities in electrodynamics can be seen to be removed in all orders of approximation. (It is these quantities which are the cause of the mathematical difficulties having to do with a lack of consistency.) In addition, the theory has not yet been extended to accommodate the phenomena that are associated with the strong, short-range type nuclear forces. However, this approach is presently in the early stages, and its discoverer feels optimistic that the approach has a good chance to become a consistent theory in the relativistic domain without the need to abandon the nondeterministic and discrete features of matter that were originally introduced with the quantum theory.

Dirac develops this approach in his recent book, LECTURES ON QUANTUM FIELD THEORY (Yeshiva, 1966).

## Other Kinds of Particles and Interactions and Quantum Field Theory

The phenomenon of radioactivity was discovered around the beginning of the twentieth century by A. H. Becquerel (1852–1908) and was investigated further by P. Curie (1859–1906) and M. Curie (1867–1934). These investigations found that a heavy substance, radium, emits three kinds of "radiation." These were called *alpha rays, beta rays,* and *gamma rays.* Later on, it was discovered that the laws of emission for these rays seemed to agree with the idea of "spontaneous decay." That is to say, such bits of "excited" radioactive matter appear to emit these three types of radiation without any outside stimulus!

Before it decays, one can consider the "excited" radioactive nucleus to be in an unstable state, just as a stone on the side of a hill would be unstable. If physical obstructions should be removed, the stone would easily roll down to the bottom of the hill. The analogy between the stone on a hill and the radioactive nucleus is not, however, a good one. The stone is, after all, interacting with the earth by means of gravitational force. The gravitational field of the earth does indeed act as an outside influence on the stone. If the earth's gravitational field were removed, the stone would not slide down to the bottom of the hill! On the other hand, the decaying radioactive nucleus does not appear to have any outside stimulus.

The *language* of quantum field theory is very appropriate to describe the *random* process of radioactive decay. It provides a means to *describe,* rather than *derive,* the creation of the alpha, beta, and gamma rays, along with the simultaneous change that is produced in the state of the radioactive nucleus. Still, it is hardly satisfactory to the theoretical physicist to rest at this stage. He is obliged to ask the question: What is the mechanism that is responsible for the "creation" of these rays?

Not too long after the discovery of radioactivity, it was found that *alpha radiation* is actually composed of many nuclei of the helium atom, which travel at high speeds. The helium nucleus, in turn, is composed of four particles that are bound together—two protons (the nuclei of hydrogen atoms) and two neutrons. The *neutron* itself was discovered in 1932 by J. Chadwick (1891–    ). The neutron is a particle that is present in all atomic nuclei, with the same or a greater abundance than the constituent protons. Its mass is almost the same as that of the proton, but unlike the proton, it has no static electric charge. On the other hand, the neutron does *bind* to other neutrons and to protons with the same strength that protons *bind* to other protons within a nucleus, such as that of helium. Since the electrical force between two protons is repulsive (that is, this force tries to keep the protons apart), and since these particles attract each other in the nucleus, one concludes that there is another type of force in operation. This is the "nuclear force" that has much greater strength, at short distances, than the electrical force. This force, which we have discussed in the preceding paragraphs in regard to the S-matrix theory, then operates in roughly the same way between protons and protons, neutrons and neutrons, and protons and neutrons.

One other very significant property that makes the neutron different from the proton is its relative "instability." When the neutron is by itself, outside of a nucleus, it appears to transform itself into a proton and some other particles. The proton, on the other hand, remains in a stable state. Even when the neutron is a constituent part of an atomic "radioactive" nucleus, it will transform in the same way, with the nucleus increasing its electrical charge by one unit (by acquiring one more constituent proton), and emitting extra particles. One of these extra particles has a negative unit of charge and was found to have the mass of an electron. The second emitted particle has no electrical charge and almost no mass.

Thus, just as in the case of the neutron, the lack of electrical charge makes it impossible for one to directly observe the latter type of particle. When the radioactive nucleus decays in this way, it is then a stream of *electrons*—the same type of particle that was discovered earlier by J. J. Thomson—that constitute the beta rays. It was the analysis of the energy distribution of these beta rays from radioactive nuclei that led to the conclusion that the emitted electrons do not take up all of the energy of the transformed neutron. The remaining energy had to reside in a second particle that could have no electrical charge, since the *uncharged* neutron decays into a proton and into an electron which each have equal magnitude but oppositely polarized electric charge. It must also have almost no inertial mass, when at rest. The latter particle was first proposed in order to explain "beta decay," by W. Pauli. It was called *neutrino*, meaning "little neutron."

The analysis of the third type of radiation from radioactive nuclei—the gamma rays—indicated that these were none other than the electromagnetic radiation that was discovered earlier in the nineteenth century and identified by Maxwell with rays of light. It was found, in this analysis, that the only difference between *gamma radiation* and radiation that corresponds to visible light is the *frequency* of the associated oscillatory electric and magnetic field intensities. Recall that when charged matter oscillates, as a pendulum does, but at thousands of times per second, radio waves are emitted. When the oscillations occur at $10^{14}$ times per second, infra-red (heat) radiation is emitted. Visible light corresponds to $10^{16}$ oscillations per second and gamma radiation corresponds to the order of $10^{23}$ oscillations per second.

To picture these oscillations, recall that electromagnetic radiation in the X ray region corresponds to electrically charged matter that would move back and forth in a box that is the size of an atom. The side of the box would be of the order of $1/10^8$ of a centimeter. Similarly, the frequency of gamma radiation corresponds to charged matter that oscillates back and forth in a box whose side is the order of the radius of an atomic nucleus. This is of the order of $1/10^{13}$ of a centimeter. While the frequency of oscillation from the radio waves to the X rays to the gamma rays changes by a large amount, the law of force that is involved here is the same in all cases. It is the law of force that is described by the equations that relate to electromagnetic phenomena.

Of the three types of processes just discussed in regard to the emitted radiation from radioactive nuclei, only the gamma radiation can be understood fairly well at the present time. Only this type of radiation has an underlying set of field equations associated with it—the Maxwell equations for electromagnetism—that give a complete accounting for the source of the observed effect. This is the case because electromagnetic forces have a long range, so that the phenomenon is easily probed. Experiments in the nineteenth century on the electric and magnetic properties of matter did, in fact, lead Faraday to the structure of the electromagnetic field theory.

On the other hand, the strong, short-range forces that govern the binding of the neutrons and protons in the atomic nuclei, and the newly discovered "weak interactions" that are responsible for the "beta decay" of neutrons into protons, electrons, and neutrinos, extend themselves over such a short range that the experiments which are necessary to deduce analogous types of field equations for these phenomena are much more difficult (and costly!) to carry out. It is a happy circumstance that, at the present time, the more affluent nations of the world are willing to donate a part of their budgets toward such experimenting. Still, the most difficult part of the problem is to cultivate the potential

Faradays and Ampères who would have the genius and the individuality to design sufficiently definitive experiments that could lead to increased understanding of these short-range forces.

High energy nuclear scattering experiments have led to the discovery of some of the properties of nuclear forces. To carry out these experiments most effectively, machines were designed to accelerate nuclear particles to extremely high energies. These experiments started in the early 1930s with the invention by E. O. Lawrence (1901–1958) of the first "atom smasher"—the *cyclotron*. Since the first machine was in operation, many more machines have been designed at increasingly higher and higher energies. Through the course of these experimental studies, it has been discovered that the "elementary particles" mentioned thus far—the photon, proton, neutron, electron, positron, and neutrino—are not the only elementary particles of matter. For a whole, new spectrum of elementary particles has been discovered which have distinct, inertial masses and are involved in seemingly distinct types of interactions. So numerous has the list of the so-called elementary particles grown to this date, that no purpose could be served, in this text, to give their specific properties. Suffice it to say at this point that the multitude of observed "elementary particles" appear, at this stage, to group together according to their masses and to the nature of their interaction with other matter.

Some of the elementary particles interact with each other, predominantly by means of the strong, short-range forces that characterize nuclear forces; some according to the weaker, long-range force that characterizes electromagnetism; and still other elementary particles interact predominantly according to the short-range (but *much weaker* than electromagnetic) force that was originally identified with beta decay. The latter are called "weak interactions." In addition to their characteristic interactions, most of the elementary particles have inertial mass values that also appear to group in a specific way.

Let us recall that the general features of inertial mass which characterize matter in motion were originally studied in an exact analytical way by Galileo and Newton, more than three hundred years ago. It wasn't, however, until the discovery of the distinct "atomic weights" by the nineteenth-century chemists, that a strong hint was given in support of the notion that matter does occur in discrete bits. The further discovery that the twentieth-century atoms—the *elementary particles*—also have a distinct set of mass values, further substantiated the notion that matter is atomistic.

It appears that the mid-twentieth-century experiments on the properties of matter would favor interpretation in terms of *discreteness* rather than *continuity* as a basic feature. Further, the success of the mathematical equations of quantum mechanics in the nonrelativistic region of energies and the seeming lack of any other continuum theory to explain these same low energy data tends to maintain the same view. While these results have given much incentive to the bulk of physics researchers to this date, the reader should not be lulled into a state of contentment. As we have emphasized earlier, there still exist in plain view some large flies in the ointment!

## Is the Space-Time Symmetry of Relativity Theory Sufficiently General?

Aside from the lack of a demonstrably consistent, mathematical description and the complete lack of any field description (in the sense of the Maxwell field theory) for the strong and the weak nuclear interactions, another persistent difficulty with the quantum theory of fields is its lack of explanation of exactly what is meant by the phrase "inertia of matter." This theory asserts that moving matter must *possess* inertia without attempting to explain its origin any further. To illustrate the latter weakness in the theory, consider the contrasting statement in which it is asserted that a piece of matter possesses $Q$ coulombs of electric charge. What this means is that such a sample of matter could be influenced by another piece of charged matter, that is, with some other quantity of intrinsic, electric charge. It would do so in a way that would cause it to move in a prescribed fashion in accordance with some exact formal description (that is, Maxwell's field equations).

On the other hand, when one says that the same piece of matter possesses $M$ units of inertial mass, does he imply that a set of equations exists which must predict the features of inertia (for example, resistance to a change in its state of motion, discrete values for the inertial masses of elementary particles, and so on)? Unfortunately, the present-day formal description of relativistic quantum field theory does not provide such a description of inertia. Finally, it might be asked: Why should one expect relativistic quantum field theory to provide such information about matter? The reason is that the latter theory claims to be a *general theory* of matter. If it is indeed to prove itself to be so, then it is obligated to contain within its potential, predictive ability, the explanation of *all* of the manifestations of matter—including its inertial properties.

It is for this reason that a great deal of attention has been given in recent years to the attempt to establish a possible relationship between the known grouping of the observed masses of elementary particles and possible *internal symmetry properties* of the equations of relativistic quantum field theory. For just as the *rotational symmetry* of the Coulomb potential, between the nucleus and the electrons, gives rise to a prediction that the solutions of the Schrödinger equation for hydrogen correspond to particular, measurable groupings of discrete values of electron energy, so an imposed hidden symmetry in the equations of quantum field theory might imply the existence of a discrete set of mass values for the elementary particles of matter. The studies of internal symmetries are also in the primitive stages at the present time and thus far leave several fundamental questions unanswered.

In spite of this, however, some quantitative answers have been obtained with this approach so that a hint does exist that it is possible that the true equations of quantum field theory are somewhat more complicated in terms of their underlying symmetry. Such a hint might then lie in the category of the approach of *axiomatic field theory* by extending the assumed axiom of space–time symmetry (that is dictated by the theory of relativity) of the equations of relativistic quantum field theory to some more complicated symmetry scheme. In view of the great amount of effort that is currently being spent on this approach, it should be established in the not-too-distant future whether or not the equations of relativistic quantum field theory need such a generalization in their underlying symmetry.

A survey and reprinting of some of the original works on symmetry in high energy physics is given in SYMMETRY GROUPS IN NUCLEAR AND PARTICLE PHYSICS, F. J. Dyson (Benjamin, 1968).

## Where Do We Stand Now on the Question of Atomism or Continuity of Matter?

We have seen that several intrinsic difficulties persist in the attempts to incorporate the quantum theory of measurement with the theory of relativity. We have also seen that a great deal of research is in progress in contemporary theoretical physics that tries to resolve these difficulties from different angles. All of these studies have one thing in common—the aim of maintaining the underlying notions of discreteness and nondeterminism that are embedded in the quantum theory of measurement, at the expense of forsaking some of the underlying notions of the completely objective, deterministic field approach that is implied by a full exploitation of the theory of relativity. While these different contemporary studies are very interesting, are complementary to each other, and could succeed in resolving the difficulties that persist, they have not yet done so! Is there still hope, then, for a continuum field theory to emerge victorious?

In answer to this question, the immediate response might be: The experimental facts do not lie! Do we not see with our very own eyes that the experiments require that the masses of the elementary particles should have a discrete set of values, rather than falling in a continuous distribution? Is it not true that the observed atomic spectra force us to say that energy, momentum, and angular momentum must have a discrete set of values, rather than lying in a continuous range, as a classical field theory would predict? Do we not have the experimental facts from the spectral distribution of blackbody radiation, the photoelectric effect, the Compton effect, the electron diffraction experiments of Davisson and Germer, and a host of other experiments in the atomic domain that force the conclusion that the values for the properties of matter in the atomic domain lie in a discrete, rather than a continuous range?

It must be admitted that the measured properties of microscopic matter appear to *approach* the feature of discreteness. In fact, by varying the experimental conditions (for example, by lowering the temperature of a gas of atoms whose radiation spectrum is being observed) one can make the observed *continuous distribution* of values come closer and closer to a discrete set of numbers. But in any measurement, it is still a continuous distribution of values that is observed—even though it *peaks* at certain places. That is, the values of the physical properties of microscopic matter, say the energy, is seen to increase very sharply at particular places on the energy spectrum, and, after reaching a maximum value, to decrease just as rapidly toward small values. The same thing happens again and again throughout the range of the energy spectrum of the observed gas.

In the quantum theory, it is *assumed* that after all of the natural effects that cause the spectrum to be continuous (but peaked) are removed, all that will be left are the discrete values of these properties. This is an assumption that can never be checked! For one of the effects that must be removed before the feature of discreteness would appear is the action of the measuring device on the system of observed atoms. Since the measurement is an integral part of the fundamental description of matter, according to the quantum theory, the idea of discreteness is an *idealization* that can, in principle, never really be observed in the actual data relating to the properties of microscopic matter. Thus, one should not overlook the possibility that the same peaked, but intrinsically continuous, distribution of values might also follow from a (nonlinear) field theory that would be based on the continuum approach, where the *limit* that is assumed in the quantum theory does not exist.

We see, then, that the most up-to-date experimental information does not compel one to adopt either the discrete or the continuum approach to a basic description of matter. Since the choice cannot be made from the actual data, it must then depend on the comparative success of the predictions that follow from each of these approaches to a theory of matter. We have seen that, to this date, the quantum approach has not been as successful as we would like, to explain both the high energy and low energy properties of matter, from a common theory. In addition to continuing research, to see if the quantum approach can eventually succeed, it is also natural to pursue the consequences of a relativistic field theory that is based on the notions of continuity and determinism. Some of the results of such an approach to unification of the predictions of the quantum and relativity theories will be discussed in the following chapter.

# 11.

# Elementary interaction field theory

Throughout the history of science, progress has usually resulted only when the scientist was willing to proceed with an open mind. So long as he insisted on one and only one approach to arrive at the solutions that were sought, he usually faced insurmountable blocks. Although great attempts were made to straddle these obstacles, which sometimes were successful, it was more usual that the obstacle could be overcome only by taking a different path, in which the obstacle did not appear from the outset.

In the previous chapter, we discussed the successes and failures of the recent, necessary attempts to unify the quantum approach to microscopic physics with the notions of relativity theory. We have seen that such attempts, in which full use has been made of the notions of the quantum theory of measurement, at the expense of forsaking some of the underlying notions of a relativistic field theory, have not yet had sufficient success to allow the quantum theory to claim itself to be a *general theory*. If it turns out that the current attempts will succeed in circumventing the intrinsic difficulties in this approach, then the intuition of a large part of the present-day physics community will be verified. It would be a triumph for the nondeterministic and discrete aspects if they are found to be basic to a true description of matter. Still, so long as the latter approach has not yet succeeded sufficiently, it is indeed in order for the objective investigator of nature to explore also the contrasting approach in which matter is described in terms of *continuous fields* with a completely determined account of all aspects of a physical system. The question that must then be answered is the following: How can one proceed, in testing this approach, to describe microscopic as well as macroscopic quantities of matter at the same time?

First, the predictions of the new theory must not disagree with those of quantum mechanics in the region of phenomena where the formal equations of the latter theory "work." It follows that if the *mathematical expression* of the new theory has a limit in the region of sufficiently low energy to neglect relativistic considerations, with precisely the same form as the equations of quantum mechanics (the Schrödinger or Heisenberg equation), then the predictions of the new theory would be precisely the same as those of quantum mechanics in the region where the latter has been successful. The two theories would then be said to "correspond" in this limit. Such a requirement of a new theory is, of course, not new. It has been adopted in almost all previous developments of new innovations that have been imposed on older theories of matter. It is the *principle of correspondence*. This is a most important principle to adhere to. It recognizes the importance of history in the development of ideas. It allows one to sift out the significant portions of our acquired knowledge upon which a higher structure of understanding must be built.

The second point is the following: If the *principle of relativity* is to be fully exploited, then it would seem that logical consistency could be maintained only if the notion that matter is made out of discrete little bits—the elementary particles—were abandoned. For if the basic constituent of matter is the *non-interacting* bit of matter (that may or may not perturb some other bit of matter at some time and at some place), then, as a *basic unit of the description,* the elementary particle would not *relate* to anything except itself. The term, relativity, *as a fundamental notion,* would then become meaningless. Thus, it is argued that the description of a physical system, to be consistent with the concept (and not only the mathematics!) that is intrinsic in the theory of relativity, must necessarily involve at least two interacting components and that their mutual interaction can never "turn off."

Of course, one must consider the cases in which the mutual coupling of the components of a system of matter becomes arbitrarily weak, as the mutual effect of two magnets might do as they are pulled apart to large separations. Nevertheless, if one should consider a magnet without its mutual influence with another magnet, the term, magnet, should lose meaning *within a theory of relativity.*

A full exploitation of the principle of relativity, according to this view, should then start with the *mutual interaction,* or said in other ways, the *mutual influence, the oneness without actual parts,* rather than the *free noninteracting thing,* as the basic entity from which to build a fundamental description of matter. Thus, instead of the elementary particle of the quantum approach, it is the elementary interaction that must here be considered as the basic object to start from.

It should be emphasized, at this point, that the demarcation between the coupled components of the elementary interaction is made in the theoretical description for convenience in consideration of the actual problem that is under study. For, when the relativistic approach is fully exploited by taking the interaction as the elementary thing, *there are not actually separated and distinct parts.* There only *appear* to be separate parts when the coupling is sufficiently weak. In the latter case, one might call one of these seemingly "separated" parts the "observer" and the other, the "observed." Nevertheless, according to the theory of relativity, it should make no difference, to the laws of nature that describe their coupling, as to which component is called "ob-

*46. Albert Einstein (1879–1955)*

server," and which is called "observed." According to this view, when the "observer" perceives its environment (the "observed") and consequently deduces the laws of nature, it should draw precisely the same conclusions about the world as would be deduced by the "observed" after it perceived its environment (the "observer"). Thus, it is the whole entity, which in fact has no distinct parts, which we call the *elementary interaction*, that must be considered as the basic building block from which a description of nature is to be constructed.

We have referred above to the elementary interaction in terms of the anthropomorphic term "observer" and "observed." This was done only for the purposes of language—indeed, for want of a better language! Since the laws of nature are contended here to be independent of which part of the system is called "observer," and which is called "observed," the description must be entirely *objective*. This is in contrast with the view of the quantum theory of measurement in which the deduced laws of nature are indeed a function of the apparatus that is doing the perceiving. The latter approach then is a *subjective* description of nature.

The previous comparison can be exemplified by considering the language structure of the sentence, "I see the *fly*." The pronoun, I, is called the subject and the noun, *fly*, is the object in the sentence. Suppose that the *I* and the *fly* should make up the entire universe. Then the sentence above appears to give a subjective description of the world. This is because the interpretation that *I* would give to what *I* am and to what a fly is, is not necessarily the same as the interpretation that the *fly* may give (no matter how primitive) to itself and to its environment (me). On the other hand, the theory of relativity asserts that the universe—the fly and me in this example—should have a fundamental underlying law of behavior that is independent of which particular part of the system (under the conditions when it is meaningful to even use the term, part) expresses the description of the universe. Thus, this theory requires that the *fly* could just as well be the subject of the sentence and *I*, the object, without changing the actual law of nature that relates to this sentence. Since the object and the subject are interchangeable in the laws of nature, according to this approach, the theory must necessarily be a completely objective description of nature. In the example, the objective entity that we have discussed is the whole unit *I–fly*. In the particle approach, on the other hand, *I* and *fly* are separate and distinct entities that may or may not perturb each other.

At this point, the reader may think that while all of this is interesting (or uninteresting), it is a bit too much on the philosophical side. After all, he might remark, the elementary particle approach of the quantum theory, as a basic description of matter, does indeed provide a set of mathematical equations that in turn predict results that are in excellent agreement with the experi-

*47. Enrico Fermi (1901–1954)*

mental facts—at least in the domain of low energy atomic physics. What is the counterpart of these equations in the elementary
interaction approach of relativity theory? True, it has already
been specified that the required field equations of the elementary
interaction theory must "correspond" with the quantum mechanical equations in the low energy (that is, nonrelativistic)
limit. But what, he may ask, is the *exact* (unapproximated) form
of these equations, how are they to be used to make predictions
about the properties of matter, and what predictions have been
made thus far in a more satisfactory way than the derivations
of the quantum mechanical equations and relativistic quantum
field theory?

While it is beyond the scope of this book to go into the

mathematical details of the elementary interaction field theory that is discussed, we will attempt, in the remainder of this chapter, to spell out some of the properties of the equations that are dictated by the physical approach.

The first property that is required of the formal description of the elementary interaction theory is that, in its simplest form, there must be at least two coupled equations to describe matter. The coupling between the component parts of the system can never be "turned off." This is because the solutions of these field equations have only to do with the manifestations of *a single interaction field*. Because of the imposed *correspondence principle*, these equations must be constructed to *approach* the form of the standard quantum mechanical equations, in the limit of no coupling. However, as we have emphasized above, the limit cannot actually be reached since the free, noninteracting thing is logically excluded from the entire approach. This is not a trivial point. It has a very serious implication in the structure of the field equations and in the predictions of the theory, predictions that do indeed differ from those that would be made with a set of equations in which the "free particle" limit would be allowed to exist.

The next point to be made, in regard to the structure of the field equations, is the following: Considering the (simplest) two-coupled equation description, the solution of one of these equations can be identified with the "observer," and the solution of the other with the "observed." However, the principle of relativity requires that it should make no difference as to which part of the single, mutual interaction is called by one of these names, and which part by the other. Thus, the equations must be unchanged in form when the solutions of the different equations are interchanged. That is, when solution *one* is called solution *two* and vice-versa, equation *one* should convert into equation *two* and equation *two* into equation *one*. Thus, no real change takes place—when the names "observer" and "observed" are interchanged; all that happens is that the fundamental field equations are rewritten in a different order, in accordance with the objective description that is required by relativity theory.

In view of the latter symmetry of the field equations, under the interchange of the solutions of the interacting "parts" of the system, it follows that equation *one* must contain solution *two* and, in precisely the same way, equation *two* must contain solution *one*. Such dependence, in fact, spells out the actual, mathematical description of the way in which one particle field influences the other. It then follows that solution *two* depends on solution *one*, and solution *one* depends on solution *two*. Thus, solution *one* depends on solution *one* itself. The fundamental equations then have the peculiar looking property that the determination of the solutions depends on an "operator" that depends on the solution itself. Such a mathematical formalism is "nonlinear" since the squares and higher powers of the solution that is sought appear in the basic equations. This fundamental *nonlinearity* in the basic field equations of the elementary interaction approach is to be contrasted with the fundamental *linearity* of the operators in the equations of quantum mechanics and quantum field theory.

To exemplify the *physical origin* of the nonlinearity of these equations, consider two interacting, electrically charged pith balls that are set into motion with respect to each other. The presence of the first pith ball in motion corresponds, according to the theory of Faraday and Maxwell, to an electromagnetic field of force which, at first, may be said to "cause" the second pith ball to move in some prescribed way. However, the resulting motion of the second pith ball also sets up a field of force, according to the same field theory, and it will thereby induce the first pith ball to change its course. We see here that the latter alteration of the path of the first pith ball ultimately resulted from its own motion! Similarly, the motion of the second pith ball also influences its own motion. Such a description is a *non-*

*linear* one. With it, one very quickly loses track of which "part" is causing which other part of the system of matter to react. In fact, the notion of a "part" seems to disappear altogether. One is then forced to the *single field of influence* as the basic entity from which to start. The "part" only appears as a mathematical limit in which the coupling between the equations becomes so weak that the system *appears* to describe "things," because they are almost separated.

In this case, the influence of the motion of the pith ball on its own motion, through the intermediary of the second pith ball, becomes negligibly small. However small this coupling may be, though, it is never gone! This feature has a profound influence on the solutions of the equations that are supposed to relate to the predictions of the theory. Thus, the *nonlinear* theory and the *linear* theory predict different things, in general, about the same physical system in the same limit of weak coupling. The experimental facts must then be used to choose which of these theories is the right one.

We see, then, that the elementary interaction approach is fundamentally a nonlinear theory. In simplified terms, this means that in such a description of matter, the *whole cannot be taken to be the simple sum of its parts.* This is in contrast with the quantum theory of elementary particles—a fundamentally linear theory of matter in which the whole is indeed the sum of its parts.

When this comparison is carried one step further, it is seen that the interpretation of the space–time coordinates must also change in a basic way that would influence the structure of the field equations, which in turn would influence the predictions of the theory. For, if matter is supposed to be composed of many elementary bits, then to be consistent with the mathematical requirement of relativity theory, each one of these bits of matter must have its own set of space–time coordinates. Thus, if a laboratory quantity of matter is made up from $N$ elementary particles, the space–time in which this matter is described must have $4N$

dimensions to represent the $N$ trajectories of the constituent elementary particles. In this description, the space and time coordinates play an *active* role; they are supposed to be the measurable properties of the parts of a system of matter.

On the other hand, the elementary interaction approach that is proposed here entails only one set of space–time coordinates; that is, we deal here with a four-dimensional space–time in which all fields are "mapped." In this picture, instead of locating the individual constituents of matter, the points of this four-dimensional space relate to the particular space and time points at which the physical strength of the interaction, for the entire system, is defined. Here, one does not assert that space and time are directly observed. Rather, they are a continuously varying set of numbers that are used as a tool in order to describe the more basic entities that relate to the observables. The latter are the continuous fields that solve the underlying equations that describe the properties of matter. But this *passive* interpretation of the space and time coordinates is not new. It is, in fact, the same interpretation of the space and time coordinates that was discussed earlier in regard to the field concept, as it was introduced by Faraday in the middle of the nineteenth century. These contrasting interpretations of space and time were indicated earlier to highlight a difference between the discrete and continuous approaches of the elementary particle and the elementary interaction theories, respectively.

To sum up, the full exploitation of the theory of relativity, through the recognition of the elementary nature of the interaction, leads to a theory of matter in terms of continuity, rather than discreteness. The elementary interaction field theory is a complete field description where the field equations are fundamentally nonlinear and where the whole entity is not describable as the sum of distinct parts. With this view, if a system of matter is enlarged to introduce more interacting matter, one cannot simply add a field to the existing description. Rather, one must start anew with the entire, enlarged system in which all field components are enmeshed and, in principle, inseparable.

The *nonlinear* field equations of the elementary interaction theory *approach* the usual, linear equations of quantum mechanics that successfully describe atomic systems in the low-energy limit. Thus, the predictions of both theories are identical in this limit, even though the underlying foundations of both theories and their *exact* mathematical expressions are quite different. This is similar to the situation that occurred when the general theory of relativity—a continuum field theory—replaced Newton's theory of universal gravitation. Newton's was a particle theory that is based on the action-at-a-distance concept to explain gravitational forces. Just as the general theory of relativity transcended the classical theory of gravitation by correctly predicting the gravitational phenomena that were not at all predicted by the classical theory, in addition to agreeing with all of the successful predictions of Newton's theory, so the elementary interaction field theory is obligated to predict, with success, more than has been predicted by the elementary particle theory of the quantum approach. This is necessary if the theory of elementary interactions is to transcend the theory of elementary particles as a more valid description of nature. In the next paragraphs, we will survey some of the successful predictions that have been made by the elementary interaction field theory and contrast them with the results of the quantum theory of fields.

Thus far, some important predictions have been made by the elementary interaction field theory that either have not yet been predicted by relativistic quantum field theory, or else are obtained by the formalism of the quantum theory, but not yet in a satisfactory way. One of the latter type of predictions relates to the energy-level structure of hydrogen—in particular, to the Lamb shift. It was indicated earlier how this effect was predicted by the quantum theory of fields, to very high precision, by utilizing a mathematical method that effectively "subtracts" the infinite quantities that appear naturally in the latter theory. The *physical model* in the latter theory involved the coupling of the bound electron in hydrogen to an infinite sea of electromagnetic radiation and electron–positron pairs (that make up the so-called "physical vacuum"). The chief difficulty in the latter derivation, however, was that the mathematical method for deriving the desired result is not demonstrably mathematically consistent.

It is not clear that other predicted values of the same measured property can also emerge from the same mathematical language without, in any way, altering the physical content of the model. On the other hand, the elementary interaction field theory, that is discussed in this chapter, has yielded numerical predictions for the Lamb shifts in the different states of the hydrogen atom that are extremely close to the measured values, but come from a *demonstrably consistent* mathematical description. In the latter theory, the result comes from the properties of a bound state of only two particles of matter—the interacting electron and proton. There is no infinite sea of radiation and a non-countable number of electron–positron pairs to make up the background "vacuum." The effect comes simply from the energy levels of the electron in the field of the proton alone.

A second important quantitative result of the field equations of the elementary interaction field theory has to do with the observations that are conventionally interpreted as the "annihilation" and "creation" of electron–positron pairs. The quantum theory of fields does not derive these effects. It rather provides a language with which to express the events of annihilation or creation of fields of matter. In a sense, though, the creation and annihilation of matter at arbitrary times and places, without the exact mechanism that causes these events to be spelled out, is a starting point for quantum field theory. The *assertion* that these events take place is an axiom of the theory and therefore need not be derived! The elementary interaction field theory, on the other hand, cannot allow matter to be created and annihilated—

let alone allow these events to occur at arbitrary times and places! This is because the annihilation of matter would automatically lead to a cessation of the interaction that represented their original presence, thereby contradicting the fundamental assertion about the elementary nature of the interaction.

Nevertheless, one does observe, in a bubble chamber or in a cloud chamber, that a pair of tracks converge to a vertex and disappear beyond it. Any theory must explain this observation! But is it really necessary to explain it in terms of the annihilation of matter? The answer is clearly, no! For what is actually observed here is the exchange of momentum and energy between the positive and negative electrically charged matter and the other matter that fills the bubble or cloud chamber. If, instead of ceasing to exist, the positively and negatively charged matter should come sufficiently close to *bind each other* in a very strong state of energy, they would then be no longer capable of giving up a part of their mutual energy to the surrounding gas in the chamber. In this case, one would no longer see the tracks that were previously identified with the "free" particle and antiparticle (that is, the electron and positron). The actual observation in this case would completely agree with this model's prediction but matter would not be annihilated here.

It also follows that if a sufficient amount of energy could be supplied to the electron–positron pair, when it is in this strongly bound state, it could become unbound (or almost so!). The separate, oppositely charged particles would once again appear to be free by separately giving up energy and momentum to the surrounding atoms of the gas in the chamber. The latter effect is also observed in practice. It is conventionally interpreted as the "creation" of matter from a vacuum. However, with this derivation from the elementary interaction theory, matter is not in fact created from a vacuum—it is only sufficiently excited from a strongly bound state so as to *appear* as separated particles of matter. But the matter was there all the time. It merely didn't communicate with the surrounding medium, before this excitation, because of an insufficient amount of available momentum transfer.

Thus far, we have only indicated the differences between what the elementary interaction theory and the elementary particle theory should predict in regard to the creation and annihilation of matter. We have indicated that the elementary particle theory does not derive the effects that are usually interpreted as creation and annihilation of matter. But, does the elementary interaction theory actually derive the strongly bound state for the electron–positron pair that must exist to explain the data without creating or annihilating matter. Indeed, one of the most interesting results of the elementary interaction field theory under discussion was the *derivation of a solution* of the underlying field equations for the particle–antiparticle pair, which exhibits all of the observed facts that are conventionally identified with the "annihilation" and "creation" of matter. But, matter is not annihilated or created here (at arbitrary times and places), as it is assumed to happen in the quantum theory of fields. The observed results are, rather, predicted from the (nonlinear) solution of a *deterministic* set of field equations in which no parameters are arbitrary.

It is important to note, in this case, that there would have been no way of constructing the nonlinear solution that gave this result from the linear solutions of the conventional equations of relativistic quantum field theory. Thus, we see here an *exact* prediction of the elementary interaction field theory which precisely duplicates a particular set of observations and has never been derived from any other theory. To complete the comparison with the observations, it was also found that the same solution exhibits a *motion* of the interacting electron and positron, in this strongly bound state, that would correspond to a measurement of two correlated currents that are oppositely polarized in a common plane. The latter effect is observed and usually identified with the "creation" of the two photons that accompany the electron–positron annihilation.

Another derivation from this formalism that is also sensitive to its *nonlinear* structure (and therefore could not have been derived from the usual quantum field theory) has to do with the *Pauli exclusion principle*. This principle, which was asserted by W. Pauli (1900–1958) in the early stages of the modern atomic theory, asserts that for a certain classification of elementary particles (including the electron and the proton), no more than one, out of a number of equivalent such particles, may be in the same place at the same time and in the same state of motion. The latter refers to the state of momentum, angular momentum, and energy. Pauli showed that this principle led to the structure of atoms that implies all of their chemical properties.

This principle also led to many more correct predictions about matter. For example, it led to important predictions about the magnetic properties of solid matter, the properties of metals (electrical conductivity, heat conductivity, binding energy, and so on), the proper way in which charged particles scatter from charged particles, the structure of nuclei, etc. Yet, the principle, while expressing an obvious truth about the low-energy properties of matter, was not *derived* from first principles, starting with the original (Dirac and Schrödinger) form of quantum mechanics. It also was not derived from the ordinary form of relativistic quantum field theory. A proof was given, however, in the development of axiomatic field theory. While the latter is indeed an important success for the quantum approach, it should be recalled that the extended theory entails the imposition of many axioms that do not yet seem to have any more than mathematical connotation and relates to a set of field equations whose mathematical consistency has not yet been established.

The elementary interaction field theory has also derived a result that is *physically equivalent* to the Pauli exclusion principle. But the only axiom that is necessary here is the *principle of relativity*, when it is fully exploited by asserting the elementary nature of the interaction. An *exact* result was obtained from the nonlinear, coupled field equations of this approach that describes an indefinitely large system of interacting components. The predicted result was that if any two of these interacting components should relate to the features that are identified with equal inertial masses, and a mutual repulsive interaction, and when they are each in the same state of motion, then the contribution of the mutual interaction of these two parts to the total interaction for the system of matter fields would automatically vanish. When this exact result of a deterministic, relativistic field theory is applied to the low-energy cases, such as atomic structure, the properties of metals, etc., the precise predictions follow which were originally summarized by Pauli in terms of his principle. This derivation was then a further success of the elementary interaction field theory, and is, therefore, in favor of the continuum and deterministic approach to the basic description of matter as it was envisioned by Faraday and Einstein.

Another important derivation of the elementary interaction field theory is in regard to the role that is played by inertial mass. In the usual elementary particle approach, inertial mass relates to an intrinsic property of the free particle of matter. But, in the elementary interaction approach, there are no "free" particles of matter. It is only the interaction that is to be treated as elementary. Thus, here the inertial mass must also relate to the mutual interaction that the moving body and the surrounding interacting matter must exert on each other. It then follows that the "mass" must not be inserted into the field equations as a parameter, to be adjusted in order to fit the data. Rather, the inertial mass must relate to a *field* of mutual interaction. Such a field does not appear in the description where all of the space–time coordinate frames are inertial (that is, where all frames of reference are in relative motion with constant velocity, or are relatively at rest). But this means, according to our discussion of relativity theory, that the variable quantity which has to do with the inertial mass of interacting matter must follow from the field properties of the space–time geometry itself. The latter is the field that expresses, in a most primitive fashion, the curvature of space–time within a Riemannian geometrical description.

To find this property of the mass of interacting matter, the field equations for matter (which have the limiting form that looks like the equations of quantum mechanics) were expressed in a curved space–time, rather than the usual "flat" space and time of Euclidean geometry. When this was done, it was found that, indeed, a field relationship does reveal itself that relates to the inertial mass of interacting matter. It was found from a property of this relationship that if space–time should, in fact, become flat, the inertial mass of any interacting matter would automatically vanish! But, according to the relationship between the curvature of space and time and the presence of matter in the universe (that is indicated in the field equations of Einstein which already successfully predicted the properties of gravitation), the results of the analysis of inertial mass, as related to the curvature of space–time, means that the inertia of interacting matter is in fact a consequence of the presence of other interacting matter in the universe. Thus, the derivation of inertial mass from the coupled field equations of the elementary interaction field theory, when it is expressed in a curved space–time, is in full agreement with the origin of inertial mass that was asserted by E. Mach in the late nineteenth century. The latter assertion was named by Einstein *Mach's principle*. The result of the present analysis indicates the remarkable (and perhaps unbelievable!) property of matter that the most inconsequential quantity (say, one electron) has a mass that depends on all of the matter in the universe!

We indicated earlier that a correct theory of matter must explain the apparent, discrete values for the inertial masses of the spectrum of elementary particles which has been seen experimentally. The field relationship between the inertial mass and the geometrical properties of space–time that was indicated in the preceding paragraph is then *obligated* to yield these mass values as a consequence of the field solutions of the equations of the theory. To date, quantitative results have not yet been obtained. However, it has been found from the analysis of the equations that in the limit, where the solutions appear to approach "free fields," there is a corresponding *approach to a discrete set of mass values*. The final quantitative results for these mass values will serve as an important test of the theory.

A further, encouraging result that followed from this derivation is the feature that the inertial mass of any amount of matter is represented by a continuous field that has the same sign under all possible experimental conditions. The implication here is that gravitational forces must always have the same sign—that is to say, gravitational forces are predicted here to be always attractive or always repulsive. To establish which of these is the fact of nature, we merely observe a single case—the attractive force of the sun on the earth. The implication is that gravitational forces must, under all possible conditions, be attractive. It also follows from a derivation within the same analysis that electromagnetic forces can be attractive or repulsive. Both of these properties of matter are in agreement with the observational facts and they have not been derived from first principles by the earlier theories.

Finally, an important feature of the elementary interaction field theory is its unification of the *inertial, electromagnetic,* and *gravitational* manifestations of interacting matter in terms of a common set of field equations that all depend on the same field variables. It is significant that the minimum description of matter, with this approach, turns out to involve the inseparability of these three properties of matter. In present studies, the attempt is being made to see if some of the solutions of this basic set of field equations might also relate to other manifestations of interacting matter such as the strong, short-range nuclear forces, the weak, short-range forces (that are responsible, for example, for beta decay), and other features of recently observed properties of matter in the very high energy region. Generally, the philosophy of the elementary interaction approach requires that *all* of the physical manifestations of interacting matter must follow from a common set of underlying field equations. These equations would, in turn, represent a law of nature that relates to a *universal interaction*. It is the view of this author that such a description is the natural generalization of the theory of general relativity.

### Summary

Since the earliest recorded history of science, there has been constant dispute between atomistic and continuous descriptions of matter, as to which are the most fundamental. Until the twentieth century, one of these approaches has always triumphed over the other because of added experimental evidence, only to be toppled again by new evidence in support of the other theory. Along the road, mankind has benefited in this constant struggle by its continual gain in understanding of nature.

The quantum theory of measurement, favoring the approach of discreteness, and the theory of relativity, favoring the approach of continuity were formulated in the twentieth century. Because of the advanced state of experimental studies of microscopic quantities of matter, it became necessary, toward the middle of the twentieth century, to fuse these two schools of thought. It turned out to be impossible to do this while maintaining the underlying premises of the two theories in totality. For as they stand, the quantum theory of measurement and the deterministic field theory that is implied by the full exploitation of the theory of relativity are both logically and mathematically inconsistent with each other. Thus, a fusion would only become possible by forsaking a part of the basis of one of these theories in favor of a full maintainance of the other.

Most of the effort in present day studies of matter is devoted to investigations that favor the quantum approach rather than the theory of relativity. The attempted fusion, known as relativistic quantum field theory, has not yet succeeded. Thus, the objective inquirer must also ask about the possible outcome of an approach that would favor the deterministic field approach of relativity theory. The results of one such study, the elementary interaction field theory, have been discussed in this chapter.

Some recent articles by this author that discuss further the elementary interaction approach and contain a bibliography of the author's mathematical articles on the subject are the following: "A New Approach to the Theory of Fundamental Processes," *Brit. Jour. Phil. Sci. 15*, 213 (1964); "The Elementarity of Measurement in General Relativity: Toward a General Theory," *Synthese 17*, 29 (1967); "On Pair Annihilation and the Einstein-Podolsky-Rosen Paradox," *Int. Jour. Theoret. Phys. 1*, 387 (1968); "Space, Time and Elementary Interactions in Relativity," *Physics Today 22*, 51 (1969).

It is the opinion of this author that because of the necessary conflict that has arisen in this century between the two *simultaneous* revolutions in science, that take opposite sides on the question of atomism versus continuity as fundamental aspects of matter, that one of these approaches will emerge a much stronger victor than has ever happened in the past. But whichever side will emerge the victor, man will have gained in increased knowledge and understanding about the world around him. He may then hope for a new conflict to arise so that his understanding may increase even further.

# FURTHER READING

Born, M., *Natural Philosophy of Cause and Chance.* (Oxford, 1948).

de Broglie, L., *The Current Interpretation of Wave Mechanics: A Critical Study* (Elsevier, 1964).

Carnap, R., *Philosophical Foundations of Physics* (Basic Books, 1966).

Eddington, A., *The Nature of the Physical World* (Macmillan, 1927).

Gamow, G., *One Two Three Infinity* (The New American Library, 1947). *Thirty Years that Shook Physics* (Doubleday, 1966).

Gershenson, D. E. and Greenberg, D. A., "The First Chapter of Aristotle's Foundations of Scientific Thought" (a translation) *The Natural Philosopher 2* (Blaisdell, 1963).

Heisenberg, W., *Physics and Philosophy* (Harper & Bros. Publ., 1958).

Hesse, M. B., *Forces and Fields* (Nelson, 1961).

Jammer, M., *The Conceptual Development of Quantum Mechanics.* (McGraw-Hill, 1966).

Kilmister, C. W., *The Environment in Modern Physics* (Elsevier, 1965).

Lanczos, C., *Albert Einstein and the Cosmic Order* (Interscience, 1965).

Mach, E., *The Science of Mechanics* (Open Court, 1960).

Munitz, M. K., editor, *Theories of the Universe* (The Free Press, 1957).

Newman, J. R., editor, *The World of Mathematics* (Simon and Shuster, 1963).

O'Rahilly, A., *Electromagnetic Theory: A Critical Examination of Fundamentals* (Dover, 1965).

Planck, M., *A Survey of Physical Theory* (Dover, 1960).

Poincaré, H., *The Foundations of Science* (The Science Press, 1929).

Popper, K. R., *Conjectures and Refutations: The Growth of Scientific Knowledge* (Harper and Row, 1965).

Prizbaum, K. and Klein, M. J., *Letters on Wave Mechanics: Einstein, Schrödinger, Planck, Lorentz* (Philosophical Library, 1967).

Russell, B., *ABC of Relativity*, 3rd edition (George Allen and Unwin, 1969).

Schilpp, P. A., editor, *Albert Einstein: Philosopher-Scientist* (Library of Living Philosophers, 1949).

Schrödinger, E., "The Meaning of Wave Mechanics", from *Louis de Broglie, Physicien et Penseur* (Éditors Albin Michel, Paris, 1952).

"The Philosophy of Experiment", *Nuovo Cimento 1*, 5 (1955).

"Are There Quantum Jumps?", *British Journal for the Philosophy of Science 3*, 3 (1952).

Tricker, R. A. R., *The Contributions of Faraday and Maxwell to Electrical Science*, (Pergamon, 1966).

# Name Index

# Subject Index

# ABOUT THE AUTHOR
# AND EDITOR

*Mendel Sachs* received his Ph.D. degree from U.C.L.A. and is now a Professor of Physics at the State University of New York in Buffalo. He was born in Portland, Oregon, and grew up in Toronto. Since then, he has taught widely and is the author of SOLID STATE THEORY in addition to numerous articles and monographs in theoretical physics.

*Daniel A. Greenberg,* consulting editor for the History of Science series, received his M.A. and Ph.D. degrees from Columbia University in 1956 and 1960, respectively. He was on the physics faculty of Columbia University and Barnard College from 1959 to 1963, and on the history faculty there from 1964 to 1966. Dr. Greenberg is presently on the staff of the Sudbury Valley School in Framingham, Massachusetts. He is the author of many articles and several books on the history of science and physics.